Modern Birkhäuser Classics

Many of the original research and survey monographs in pure and applied mathematics published by Birkhäuser in recent decades have been groundbreaking and have come to be regarded as foundational to the subject. Through the MBC Series, a select number of these modern classics, entirely uncorrected, are being re-released in paperback (and as eBooks) to ensure that these treasures remain accessible to new generations of students, scholars, and researchers.

T0178051

Fourier Integral Operators

J.J. Duistermaat

Reprint of the 1996 Edition

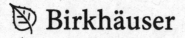
Birkhäuser

J.J. Duistermaat (deceased)

Originally published as Volume 130 in the series *Progress in Mathematics*

ISBN 978-0-8176-8107-4   e-ISBN 978-0-8176-8108-1
DOI 10.1007/978-0-8176-8108-1
Springer New York Dordrecht Heidelberg London

www.birkhauser-science.com

J. J. Duistermaat

Fourier Integral Operators

Birkhäuser

Boston • Basel • Berlin

J. J. Duistermaat
Mathematics Institute
Universiteit Utrecht
3508 TA Utrecht
The Netherlands

Library of Congress Cataloging-in-Publication Data

Duistermaat, J. J. (Johannes Jisse), 1942-
    Fourier integral operators / J. J. Duistermaat
        p.   cm. -- (Progress in mathematics ; v. 130)
    Includes bibliographical references.
    ISBN 0-8176-3821-0 (h : acid free). -- ISBN 3-7643-3821-0 (h :
acid free)
        1. Fourier integral operators.   2.   Fourier series.   I. Title.
II. Series: Progress in mathematics (Boston, Mass.) ; vol. 130.
QA329.6.D843 1995   95-35173
515'.723--dc20   CIP

Printed on acid-free paper   *Birkhäuser* 🏦®

© Birkhäuser Boston 1996

ISBN 0-8176-3821-0
ISBN 3-7643-3821-0

Layout and typesetting by Martin Stock, Cambridge, MA
Printed and bound by Quinn-Woodbine, Woodbine, NJ
Printed in the U.S.A.

9 8 7 6 5 4 3 2 1

Contents

Preface

More than twenty years ago I gave a course on Fourier Integral Operators at the Catholic University of Nijmegen (1970-71) from which a set of lecture notes were written up; the Courant Institute of Mathematical Sciences in New York distributed these notes for many years, but they became increasingly difficult to obtain. The current text is essentially a nicely TeXed version of those notes with some minor additions (e.g., figures) and corrections.

Apparently an attractive aspect of our approach to Fourier Integral Operators was its introduction to symplectic differential geometry, the basic facts of which are needed for making the step from the local definitions to the global calculus. A first example of the latter is the definition of the wave front set of a distribution in terms of testing with oscillatory functions. This is obviously coordinate-invariant and automatically realizes the wave front set as a subset of the cotangent bundle, the symplectic manifold in which the global calculus takes place.

Similarly, the principal symbol of a Fourier integral distribution is defined as the leading term in the asymptotic expansion of testing with oscillatory functions. In this way the principal symbol is identified with a certain function on a space of Lagrange planes. This leads to a definition of the Maslov line bundle which looks somewhat different from the usual one, but which fits naturally with the idea of characterizing singularities of distributions by testing with oscillatory functions. It should also be noted that the asymptotic expansion is obtained by applying the method of stationary phase, which is the central analytical tool in the theory.

The text contains two basic applications of the theory: the Cauchy problem for strictly hyperbolic equations and caustics in oscillatory integrals. We have not attempted to treat the numerous other applications which have been developed since the seventies, because that would change the book from an introduction into a research monograph. For this we refer the reader to the excellent four volume book by Hörmander, *The Analysis of Linear Partial Differential Operators* published by Springer-Verlag, of

which the Notes at the end of each chapter can be used to get quite a
complete overview of the applications.

J. J. Duistermaat
September, 1995

Fourier Integral Operators

Chapter 0

Introduction

Let $\mathcal{S}$ be the Schwartz space of all complex valued C^∞ functions u on $\mathbb{R}^n$ such that $x^\beta(\frac{\partial}{\partial x})^\alpha u$ is bounded on $\mathbb{R}^n$ for all α, β. Here $x = (x_1, \ldots, x_n) \in \mathbb{R}^n$, $\alpha = (\alpha_1, \ldots, \alpha_n)$, $\beta = (\beta_1, \ldots, \beta_n)$ are n-tuples of non-negative integers,

$$(0.1) \qquad x^\beta = \prod_{j=1}^n x_j^{\beta_j}, \qquad \left(\frac{\partial}{\partial x}\right)^\alpha = \prod_{j=1}^n \frac{\partial^{\alpha_j}}{\partial x_j^{\alpha_j}}.$$

As is well known, the *Fourier transformation*, defined by

$$(0.2) \qquad (\mathcal{F}u)(\xi) = \int e^{-i\langle x, \xi\rangle} u(x)\, dx$$

yields an isomorphism: $\mathcal{S} \to \mathcal{S}$, with inverse given by

$$(0.3) \qquad (\mathcal{F}^{-1}v)(x) = (2\pi)^{-n} \int e^{i\langle x, \xi\rangle} v(\xi)\, d\xi.$$

Here $x, \xi \in \mathbb{R}^n$ and

$$(0.3a) \qquad \langle x, \xi\rangle = \sum_{j=1}^n x_j \xi_j$$

denotes the inner product between the vector x and the covector ξ.

In particular, we can write the identity as:

$$(0.4) \qquad u(x) = (2\pi)^{-n} \iint e^{i\langle x-y, \xi\rangle} u(y)\, dy\, d\xi.$$

Now consider a linear partial differential operator

$$(0.5) \qquad P = \sum_{|\alpha| \leq m} a_\alpha(x)\left(\frac{\partial}{\partial x}\right)^\alpha$$

with variable coefficients $a_\alpha(x)$ and of order m. ($|\alpha| = \alpha_1 + \ldots + \alpha_n$ denotes the *order* of $(\frac{\partial}{\partial x})^\alpha$.) Then we obtain from (0.4):

$$(0.6) \qquad (Pu)(x) = (2\pi)^{-n} \iint e^{i\langle x-y, \xi\rangle} \sigma_P(x, \xi) u(y)\, dy\, d\xi,$$

P. Buser, *Fourier Integral Operators*, Modern Birkhäuser Classics,
DOI 10.1007/978-0-8176-8108-1_1, © Springer Science+Business Media, LLC 2011

where the *symbol* σ_P of P is given by

(0.7) $$\sigma_P(x,\xi) = \sum_{|\alpha| \le m} a_\alpha(x)\,(i\xi)^\alpha.$$

So σ_P is a polynomial in ξ of degree m with coefficients depending on x.

If P has constant coefficients then one can try to find an inverse E for P by formally writing

(0.8) $$(Ef)(x) = (2\pi)^{-n} \iint e^{i\langle x-y,\xi\rangle} \sigma_P(\xi)^{-1} f(y)\,dy\,d\xi.$$

Of course the possible zeros of σ_P in $\mathbb{R}^n$ and decrease of σ_P at infinity in general will cause trouble. However by pushing the integration into the complex domain $\mathbb{C}^n$ in a suitable way and applying Cauchy's integral formula (here one uses that σ_P is analytic), the zeros can be avoided. In this way one can obtain a *(right) fundamental solution* for P, that is, a continuous linear map $E : C_0^\infty(\mathbb{R}^n) \to C^\infty(\mathbb{R}^n)$ that can be extended to a continuous linear map $\mathcal{E}'(\mathbb{R}^n) \to \mathcal{D}'(\mathbb{R}^n)$ such that $PEf = f$ for all $f \in \mathcal{E}'(\mathbb{R}^n)$. (Usually the distribution $E\delta$ is called the fundamental solution.) For a more detailed treatment of this result, which is due to Ehrenpreis [26] and Malgrange [60]; see Hörmander [44], Ch. III or [42'], Ch. X.

The above procedure breaks down almost completely if we allow P to have variable (but still smooth, or even analytic) coefficients. However, for an important special class of operators there is a rather satisfactory substitute. In order to describe this, we start with some definitions.

The *principal part p of the symbol of P* is defined by

(0.9) $$p(x,\xi) = \sum_{|\alpha|=m} a_\alpha(x)\,(i\xi)^\alpha,$$

that is, the homogeneous part of highest order of σ_P.

The operator P is now called *elliptic* if

(0.10) $$\xi \ne 0 \quad \text{implies} \quad p(x,\xi) \ne 0.$$

The best-known examples of elliptic operators are

(i) all nonsingular ordinary differential operators (that is, operators acting on $\mathbb{R}^1$),

(ii) the Cauchy–Riemann operator $\frac{1}{2}\left(\frac{\partial}{\partial x_1} + i\frac{\partial}{\partial x_2}\right)$ on $\mathbb{R}^2$,

(iii) the Laplacian $\Delta = \sum_{j=1}^{n} \frac{\partial^2}{\partial x_j^2}$ in $\mathbb{R}^n$.

Note that the wave operator $\frac{1}{c^2}\frac{\partial^2}{\partial t^2} - \Delta$ is not elliptic. Of course, it is easy to vary (ii), (iii) to operators with variable coefficients.

A *pseudodifferential operator A of order* μ is, in its simplest form, given by

$$(0.11) \qquad (Au)(x) = (2\pi)^{-n} \iint e^{i\langle x-y,\xi\rangle} a(x,\xi)\, u(y)\, dy\, d\xi,$$

where the *symbol* $a(x,\xi)$ of the operator A is a C^∞ function on $\mathbb{R}^n \times \mathbb{R}^n$ and satisfies an asymptotic development of the form

$$(0.12) \qquad a(x,\xi) \sim \sum_{j=0}^{\infty} a_j(x,\xi).$$

Here the a_j are C^∞ functions defined on $\mathbb{R}^n \times (\mathbb{R}^n \setminus \{0\})$ and positively homogeneous of degree $\mu - j$, that is,

$$(0.13) \qquad a_j(x,\tau\xi) = \tau^{\mu-j} a_j(x,\xi) \quad \text{if} \quad \tau > 0.$$

The asymptotic development (0.12) means that

$$(0.14) \qquad a(x,\xi) - \sum_{j=0}^{k-1} a_j(x,\xi) = O(|\xi|^{\mu-k}) \quad \text{for} \quad |\xi| \to \infty$$

(and similar estimates for the derivatives, see Section 2.1). The leading term $a_0(x,\xi)$ is called the *principal symbol* of the operator A.

Of course each partial differential operator is also a pseudodifferential operator with symbol equal to σ_P, so we have a true generalization of the class of differential operators. (It is important that the order μ may be any real number.) If P is a differential operator of order m, with principal symbol p, and A is pseudodifferential of order μ, with principal symbol a_0, then PA is pseudodifferential of order $m+\mu$ and with principal symbol $p{\cdot}a_0$, as is easily verified. (In Sec. 2.5 we shall give the more complete calculus of pseudodifferential operators as developed by Calderón–Zygmund [12], Kohn–Nirenberg [50] and Hörmander [42] and [40], Ch. 2.)

Now let P be an elliptic differential operator of order m, with principal symbol p. Choose any pseudodifferential operator $A^{(0)}$ of order $-m$ with principal symbol $a_0^{(0)} = 1/p$. Then $R^{(1)} = I - PA^{(0)}$ is pseudodifferential of order -1 with principal symbol, say r_1. Let $A^{(1)}$ be a pseudodifferential operator of order $-m-1$ with principal symbol $a_0^{(1)} = r_1/p$. Then $R^{(2)} = I - P(A^{(0)} + A^{(1)})$ is of order -2. Going on like this we obtain a sequence

$A^{(j)}$, $j = 0, 1, 2, \ldots$ of pseudodifferential operators of order $-m - j$ such that

$$(0.15) \qquad R^{(k+1)} = I - P\left(A^{(0)} + \cdots + A^{(k)}\right)$$

is of order $-(k + 1)$. Let $a^{(j)}$ be the complete symbol of $A^{(j)}$ and let A be a pseudodifferential operator with a symbol a such that

$$(0.16) \qquad a \sim \sum_{j=0}^{\infty} a^{(j)}.$$

(The existence of such an a requires a little additional work.) Then $R = I - PA$ is a pseudodifferential operator with symbol r such that

$$(0.17) \qquad r(x, \xi) = O(|\xi|^{-k}) \text{ for } |\xi| \to \infty, \text{ any } k.$$

It follows immediately that the integral

$$(0.18) \qquad K(x, y) = (2\pi)^{-n} \int e^{i\langle x - y, \xi\rangle} r(x, \xi)\, d\xi$$

converges absolutely and differentiation under the integration sign shows that it defines a C^{∞} function. So

$$(0.19) \qquad (Ru)(x) = \int K(x, y) u(y)\, dy$$

and we conclude that R is an integral operator with C^{∞} kernel. In other words. we have solved the equation $PA = I$ with a pseudodifferential operator A, but only modulo an integral operator with smooth kernel. Such an approximate solution of the equation $PA = I$ is called a *right parametrix* for the operator P.

For several purposes a right parametrix is just as good as a right fundamental solution. For instance, considering R as an operator: $L^2(U) \to L^2(U)$, U an open neighborhood of x, we see that R has an operator norm as small as we want by taking U small. But then the "Neumann series"

$$(0.20) \qquad (I - R)^{-1} = \sum_{k=0}^{\infty} R^k$$

converges in the space of continuous linear operators in $L^2(U)$ and we find for every $f \in L^2(U)$ a solution $u = A(I - R)^{-1}f$ such that $Pu = f$ in U. So the existence of a right parametrix implies *local solvability* of the equation $Pu = f$. (More generally, Riesz theory shows that for any relatively compact $U \subset \mathbb{R}^n$ the equation $Pu = f$ is solvable in U for all f that are orthogonal to some finite-dimensional subspace of $C_0^\infty(U)$.) A precise description of the *singularities* of Af, in terms of the singularities of f, will be given later on for any pseudodifferential operator A, and this will lead to the famous theorem that u can only have singularities where Pu has singularities, if P is elliptic. So the parametrix will also give information on the regularity of the solutions of an equation $Pu = f$.

We now turn to an example showing that the solutions of genuinely nonelliptic problems sometimes can be obtained by application of an operator that is not pseudodifferential, but still admits an integral representation resembling (0.11). Consider the initial value problem for the wave equation:

$$(0.21) \qquad \frac{1}{c^2}\frac{\partial^2 u}{\partial t^2} - \Delta u = 0$$

$$(0.22) \qquad u(x,0) = u_0(x), \qquad \frac{\partial u}{\partial t}(x,0) = u_1(x).$$

Introduce

$$(0.23) \qquad \hat{u}(\xi,t) = \int e^{-i\langle x,\xi\rangle} u(x,t)\, dx,$$

that is, the Fourier transform of u only with respect to the space variables. Then (0.21), (0.22) turn into the ordinary differential equation

$$(0.24) \qquad \frac{\partial^2 \hat{u}}{\partial t^2} + c^2|\xi|^2\hat{u} = 0$$

with initial conditions

$$(0.25) \qquad \hat{u}(\xi,0) = \hat{u}_0(\xi), \qquad \frac{\partial\hat{u}(\xi,t)}{\partial t} = \hat{u}_1(\xi),$$

all depending on the parameter ξ. This leads to the solution formula (cf. Cauchy [15]):

$$(0.26) \qquad u(x,t) = (2\pi)^{-n}\iint e^{i(\langle x-y,\xi\rangle \pm c|\xi|t)} \cdot \tfrac{1}{2}(u_0(y) \pm u_1(y)/ic|\xi|)\, dy\, d\xi,$$

where $\pm$ means taking the sum of two terms, one with the $+$ sign and the other with the $-$ sign everywhere. This resembles (0.11), however with the phase function $\langle x - y, \xi\rangle$ replaced by $\langle x - y, \xi\rangle \pm c|\xi|t$. So we have now arrived at a general type of "Fourier integral operator" of the form

$$(0.27) \qquad (Au)(x) = \iint e^{i\varphi(x,y,\xi)} a(x,y,\xi) u(y)\, dy\, d\xi,$$

where the "phase function" φ may be rather general. However, it will still be required to be real and homogeneous of degree 1 (that is,

$$\varphi(x, y, \tau\xi) = \tau\,\varphi(x, y, \xi) \text{ if } \tau > 0).$$

We also retain the condition that the "amplitude function" a has an asymptotic development of the form (0.12).

Functions of the form $e^{i\tau\varphi(x)}a(x)$ were introduced perhaps first in a remark of Debye in [77] in order to explain why geometrical optics is such a good approximation for the behavior of light. Note that since Maxwell [63], III, Ch. XX, light was generally accepted to be just electromagnetic waves. So how do we get from the partial differential equations governing these to the much simpler geometrical optics? The idea is as follows. Let the partial differential operator P (with variable coefficients if we want to treat nonhomogeneous media such as lenses!) have real principal symbol $p(x, \xi)$. Then $e^{-i\tau\varphi}P(e^{i\tau\varphi}a)$ will be a polynomial of degree m in τ, with highest-order term equal to $\tau^m\, p\big(x, \frac{\partial\varphi}{\partial x}(x)\big) \cdot a(x)$ and next-highest order term equal to

$$\tau^{m-1}\frac{1}{i}\left(\sum_{j=1}^{n} \frac{\partial p}{\partial \xi_j}\Big(x, \frac{\partial\varphi(x)}{\partial x}\Big) \cdot \frac{\partial a}{\partial x_j} + q \cdot a\right),$$

with q some function of x, depending on P and φ. So if we first choose φ such that

$$(0.28) \qquad p\Big(x, \frac{\partial\varphi(x)}{\partial x}\Big) = 0$$

and then take $a \neq 0$ equal to a solution of

$$(0.29) \qquad \sum_{j=1}^{n} \frac{\partial p}{\partial \xi_j}\Big(x, \frac{\partial\varphi(x)}{\partial x}\Big) \cdot \frac{\partial a}{\partial x_j} + q \cdot a = 0$$

then we have that $P(e^{i\tau\varphi}a) = O(\tau^{m-2})$ as $\tau \to \infty$. Such functions $e^{i\tau\varphi}a$ will be called *asymptotic oscillatory solutions*. In the case of the wave equation (0.28) turns out to be the *eikonal equation* of geometric optics, whereas the *transport equation* (0.29) describes the growth of the amplitude along the orbits of the dynamical system

$$(0.30) \qquad \frac{dx}{dt} = \frac{\partial p}{\partial \xi}\Big(x, \frac{\partial\varphi(x)}{\partial x}\Big).$$

(A similar device is used in quantum theory, in connection with the "WKB-method".) Note that (0.28) is a *nonlinear* first-order partial differential equation for φ. A detailed treatment of such equations will be given in Ch. 3, Section 7. The transport equation (0.29) is much simpler, since it can be reduced to the ordinary differential equation $\frac{d}{dt}a(x(t)) + q(x(t)) \cdot a(x(t)) = 0$, with $x(t)$ a solution of (0.30).

Birkhoff [9] introduced the refinement of replacing $a(x)$ by

$$(0.31) \qquad a(x,\tau) \sim \sum_{j=0}^{\infty} a_j(x)\,\tau^{\mu-j},$$

proving that the a_j can be chosen recursively in such a way that $P(e^{i\tau\varphi}a) = O(\tau^{-k})$ as $\tau \to \infty$ for all k. (Exercise!)

In the case of systems such as the Maxwell equations the method applies equally well. The equations for the leading term $a_0(x)$, now vector valued, reveal such interesting features as polarization and energy transport along light rays. See Luneburg [59] and the book of Kline and Kay [49]. (See also Born and Wolf [11], p. 109 for historical remarks.) Finally, Lax [53], Courant and Lax [19] showed that integrals (0.27) involving the asymptotic solutions $e^{i\varphi}a$ can be used to represent the solutions of general hyperbolic equations, leading to a proof of the generalized Huygens principle for such equations. (For a slightly more general setup, see Ludwig [56].) This will be treated in Section 5.2.

A general theory of Fourier integral operators (0.27) was given by Hörmander [40], including an invariant definition of a "principal symbol" of such operators. (The problem is that one operator may be represented by different phase functions φ_1 and φ_2, and corresponding amplitudes a_1 and a_2, respectively, and the idea is to give a definition of the principal symbol that does not depend on such a choice. A similar concept is contained in the construction of the "canonical operator" of Maslov [61], Part 2, Ch. 2.) The reader will find that we have dwelt quite extensively on symplectic geometry, which serves here as a preliminary to the invariant theory of Fourier integral operators. However this subject has much interest of its own and is basic for the understanding of nonlinear first-order partial differential equations, variational calculus and classical mechanics. For these reasons I have tried to give in Ch. 3 a rather complete account of this classical theory.

Chapter 1

Preliminaries

1.1. Distribution densities on manifolds

We assume the basic concepts of distribution theory (as for instance in Hörmander [44], Ch. 1), manifolds and vector bundles to be known.

Let E be an n-dimensional vector space over $\mathbb{R}$, $\Lambda^n E$ the space of n-vectors in E, defined for instance as the dual of the space of n-linear alternating forms: $E^n \to \mathbb{R}$. $\Lambda^n E$ is 1-dimensional over $\mathbb{R}$. For any $\alpha \in \mathbb{R}$ we call a *complex valued density* (or *nonoriented volume*) *of order* α each mapping $\rho: \Lambda^n E \setminus \{0\} \to \mathbb{C}$ such that $\rho(\lambda v) = |\lambda|^\alpha \cdot \rho(v)$ for each $v \in \Lambda^n E \setminus \{0\}$, $\lambda \in \mathbb{R} \setminus \{0\}$. The space of all densities of order α is 1-dimensional over $\mathbb{C}$ and will be denoted by $\Omega_\alpha(E)$.

Now let X be an n-dimensional C^∞ manifold. The tangent space of X at $x \in X$ will be denoted by $T_x(X)$. The $\Omega_\alpha(T_x(X))$, $x \in X$ are the fibers of a C^∞ complex line bundle $\Omega_\alpha(X)$ over X in a natural way. A C^∞ *density on X of order* α is now defined as a C^∞ section $\rho: X \to \Omega_\alpha(X)$. The space of C^∞ densities of order α on X will be denoted by $C^\infty(X, \Omega_\alpha)$. Note that after the choice of a nowhere vanishing standard density of order α, the space $C^\infty(X, \Omega_\alpha)$ can be identified with $C^\infty(X)$. Using a partition of unity one can always construct a strictly positive C^∞ density of order α on X if X is paracompact (as will always be assumed for C^∞ manifolds here).

If E, F are n-dimensional vector spaces over $\mathbb{R}$, then an injective linear mapping $A: E \to F$ induces the mapping $A_*: \Lambda^n E \to \Lambda^n F$ and so induces a mapping $A^*: \Omega_\alpha(F) \to \Omega_\alpha(E)$ defined by

(1.1.1) $$(A^* \rho)(v) = \rho(A_* v), \qquad v \in \Lambda^n E \setminus \{0\},$$

and called *pull-back of densities of order* α. If X, Y are n-dimensional C^∞ manifolds and Φ is a C^∞ immersion: $X \to Y$ then we define the pull-back $\Phi^*: C^\infty(Y, \Omega_\alpha) \to C^\infty(X, \Omega_\alpha)$ by $(\Phi^* \rho)(x) = (D\Phi_x)^*(\rho(\Phi(x)))$. Here $D\Phi_x$ is the differential of the mapping Φ, which is a linear mapping:

P. Buser, *Fourier Integral Operators*, Modern Birkhäuser Classics,
DOI 10.1007/978-0-8176-8108-1_2, © Springer Science+Business Media, LLC 2011

$T_x(X) \to T_{\Phi(x)}(Y)$. (That Φ is an immersion means precisely that $D\Phi_x$ is injective for all $x \in X$.) If Φ is a diffeomorphism: $X \to Y$ then it induces a mapping $\tilde{\Phi}: \Omega_\alpha(Y) \to \Omega_\alpha(X)$ defined by

$$(1.1.2) \qquad \tilde{\Phi}(y, \rho) = (\Phi^{-1}(y), D\Phi_x^*(\rho)), \quad y \in Y, \, \rho \in \Omega_\alpha(T_y(Y)).$$

(If $E \xrightarrow{\pi} X$ is a bundle over X then a point $e \in \pi^{-1}(x)$ is denoted by (x, e) rather than by e, in order to keep in mind in which fiber of the bundle the point e is lying.) In fact the mappings induced by the local coordinatizations of X are used to make $\Omega_\alpha(X)$ into a C^∞ complex line bundle.

If $E = F$ then $A_*: \Lambda^n E \to \Lambda^n E$ is equal to multiplication with $\det A$, so $A^*: \Omega_\alpha(E) \to \Omega_\alpha(E)$ is equal to multiplication with $|\det A|^\alpha$. So we see that for every $\rho \in C^\infty(X, \Omega_\alpha)$ the $\rho_\kappa = (\kappa^*)^{-1}(\rho) \in C^\infty(\kappa(u))$, κ a local coordinatization: $U \to \mathbb{R}^n$, form a collection of functions with transformation formula:

$$(1.1.3) \qquad \rho_{\kappa_2} = \left(\rho_{\kappa_1} \circ (\kappa_1 \circ \kappa_2^{-1})\right) \cdot \left|\det D(\kappa_1 \circ \kappa_2^{-1})\right|^\alpha.$$

We could also have started by defining a density of order α as a collection of functions ρ_κ, κ local coordinatization of X, satisfying (1.1.3), but we preferred to start with a coordinate free description of the line bundle $\Omega_\alpha(X)$ of which the densities of order α are sections.

Note that densities of order 0 are just complex valued functions on X. On the other hand densities of order 1 are the densities on X in the usual sense. If $\rho \in C^\infty(X, \Omega_1)$ has compact support, then there is a coordinate invariant *integral* of ρ over X, denoted by $\int \rho \, dx$. Here the *support* of ρ, denoted by supp ρ, is defined as the closure in X of the set of $x \in X$, such that $\rho(x) \neq 0$.

If $\rho \in C^\infty(X, \Omega_\alpha)$, $\sigma \in C^\infty(X, \Omega_\beta)$ then pointwise multiplication leads to a product $\rho \cdot \sigma \in C^\infty(X, \Omega_{\alpha+\beta})$. In particular

$$(\rho, \sigma) \to \int \rho \cdot \sigma \, dx$$

defines a continuous bilinear form on $C(X, \Omega_\alpha) \times C_0^\infty(X, \Omega_{1-\alpha})$ so $\sigma \to \int(\rho \cdot \sigma) \, dx$ is an element of $(C_0^\infty(X, \Omega_{1-\alpha}))'$ that will also be denoted by ρ. It follows that we have a continuous embedding: $C^\infty(X, \Omega_\alpha) \to (C_0^\infty(X, \Omega_{1-\alpha}))'$, and for this reason $(C_0^\infty(X, \Omega_{1-\alpha}))'$ is called the space $\mathcal{D}'(X, \Omega_\alpha)$ of *distribution densities of order* α. (Distribution densities were introduced first by de Rham [20].)

A special case arises when $1 - \alpha = \alpha$, that is, $\alpha = 1/2$. In this case we have a natural duality between $C^\infty(X, \Omega_{1/2})$ and $C_0^\infty(X, \Omega_{1/2})$, and $\mathcal{D}'(X, \Omega_{1/2})$ is equal to the dual space of $C_0^\infty(X, \Omega_{1/2})$. This simplification is also convenient if applied in:

Theorem 1.1.1 (Schwartz [74]). *Let X, Y be paracompact manifolds. Then the formula*

$$(1.1.4) \quad (A\psi)(\varphi) = a(\varphi \otimes \psi), \qquad \varphi \in C_0^\infty(X, \Omega_\alpha), \quad \psi \in C_0^\infty(Y, \Omega_\alpha),$$

defines a bijective relation between the continuous linear mappings A: $C_0^\infty(Y, \Omega_\alpha) \to \mathcal{D}'(X, \Omega_{1-\alpha})$ and the distributions $a \in \mathcal{D}'(X \times Y, \Omega_{1-\alpha})$ on the product space $X \times Y$. Here the tensor product $\varphi \otimes \psi$ is defined by

$$(1.1.5) \qquad\qquad (\varphi \otimes \psi)(x, y) = \varphi(x) \cdot \psi(y).$$

That every $a \in \mathcal{D}'(X \times Y, \Omega_{1-\alpha})$ leads to a continuous linear operator $A: C_0^\infty(Y, \Omega_\alpha) \to \mathcal{D}'(X, \Omega_{1-\alpha})$ is easy to see, and in fact we will only use Theorem 1.1.1 in this way. The converse, which is the hard part, states that every continuous linear operator $A: C_0^\infty(Y, \Omega_\alpha) \to \mathcal{D}'(X, \Omega_{1-\alpha})$ arises in this way. The distribution a is called the *distribution kernel* of the operator A.

1.2. The method of stationary phase

In this section we investigate the asymptotic behavior of integrals of the form

$$(1.2.1) \qquad\qquad I(a, t) = \int e^{itf(x,a)} g(x, a, t) \, dx$$

for $t \to \infty$. Here the integration is over $x \in \mathbb{R}^n$, a is a parameter varying in $\mathbb{R}^p$. The "phase function" f is assumed to be real-valued and smooth on $\mathbb{R}^n \times \mathbb{R}^p$, and for the amplitude function g we assume that

$$(1.2.2) \quad g \in C^\infty(\mathbb{R}^n \times \mathbb{R}^p \times \mathbb{R}^+), \qquad g(x, a, t) = 0 \text{ for } x \notin K, \, a \in A.$$

Here K, respectively A, is some fixed compact subset of $\mathbb{R}^n$, respectively $\mathbb{R}^p$. Moreover,

$$(1.2.3) \qquad\qquad \left(\frac{\partial}{\partial x}\right)^\alpha g = O(t^{m+\delta|\alpha|}) \text{ for } t \to \infty,$$

uniformly in $(x, a) \in K \times A$.

Proposition 2.1.1. *Assume that $\delta < 1$ in (1.2.3). Write*

$$(1.2.4) \qquad \Sigma_f = \{(x,a) \in K \times A;\ d_x f(x,a) = 0\}$$

for the set of stationary points of f with respect to the integration variables. Assume that for every N there exists a neighborhood Ω of Σ_f in $K \times A$ such that

$$(1.2.5) \qquad g(x,a,t) = O(t^{-N}) \text{ for } t \to \infty, \text{ uniformly in } (x,a) \in \Omega.$$

Then we have, for each N,

$$(1.2.6) \qquad I(a,t) = O(t^{-N}) \text{ for } t \to \infty, \text{ uniformly in } a \in A.$$

Proof. Using a cutoff function for Σ_f, we see in view of (1.2.5) that we only need to consider the case that $d_x f(x,a) \neq 0$ for all $(x,a) \in (K \times A) \cap \operatorname{supp} g$. Now let $L = \Sigma c_j(x,a)\frac{\partial}{\partial x_j}$, $c_j \in C^\infty(\mathbb{R}^n \times \mathbb{R}^p)$ be such that

$$Lf = 1 \text{ in a neighborhood of } (K \times A) \cap \operatorname{supp} g.$$

This equation can easily be satisfied locally; a partition of unity then leads to a global solution L in a neighborhood of $(K \times A) \cap \operatorname{supp} g$. But then

$$I(a,t) = \int e^{itf} g\, dx = (it)^{-1}\int L(e^{itf}) \cdot g\, dx$$

$$= (it)^{-1}\int e^{itf} \cdot L'g\, dx,$$

and repeating this procedure we obtain

$$(1.2.7) \qquad I(a,t) = (it)^{-k}\int e^{itf} \cdot (L')^k g\, dx$$

for all k. Here L' denotes the transposed operator of L with respect to the x-variables. It follows in view of (1.2.3) that $I(a,t) = O(t^{m+\delta k - k})$ for $t \to \infty$ and for all k, so (1.2.6) holds if $\delta < 1$. $\qquad\square$

Proposition 2.1.1 shows that in the study of (1.2.1) we can concentrate on the set Σ_f of points where the phase function f is stationary with respect to the integration variables. We now investigate the asymptotic behavior of (1.2.1) not assuming that g is rapidly decreasing near Σ_f. Instead we make a rather strong assumption on the behavior of f, namely:

$$(1.2.8) \qquad \begin{array}{l} d_x^2 f(x,a) \text{ is a nondegenerate quadratic form} \\ \text{if } x \in K,\ a \in A,\ d_x f(x,a) = 0. \end{array}$$

In this case we have the following Morse lemma with parameters.

Lemma 1.2.2. *Let $f(x,a)$ be a C^∞ function in a neighborhood of $(0,0)$ in $\mathbb{R}^n \times \mathbb{R}^p$ such that $d_x f(0,0) = 0$ and $d_x^2 f(0,0)$ is nondegenerate.*

Then there exists a neighborhood A of 0 in $\mathbb{R}^p$ and a neighborhood X of 0 in $\mathbb{R}^n$ such that for each $a \in A$ there is exactly one $x = x(a) \in X$ satisfying $d_x f(x,a) = 0$; $a \to x(a)$ is C^∞ from A into X.

Furthermore there exists a C^∞ mapping $y\colon X \times A \to \mathbb{R}^n$ such that

$$(1.2.9) \qquad y(x,a) = x - x(a) + O(|x - x(a)|^2),$$

and

$$(1.2.10) \quad f(x,a) = f(x(a),a) + \tfrac{1}{2}\langle Q(a)y, y\rangle, \qquad Q(a) = d_x^2 f(x(a),a).$$

Proof. The first part is a direct application of the implicit function theorem. By transforming to the new variable $x - x(a)$ we may now assume that $d_x f(0,a) = 0$.

Try $y(x,a) = R(x,a)x$, with $R(x,a)$ an $n \times n$ matrix, $R(0,a) = I$. So we want $f(x,a) - f(0,a) = \tfrac{1}{2}\langle Q(a)Rx, Rx\rangle$. Applying Taylor expansion of $t \mapsto f(tx,a)$ of order two, with integral remainder term, we obtain

$$f(x,a) - f(0,a) = \tfrac{1}{2}\langle B(x,a)x, x\rangle,$$

where

$$\tfrac{1}{2}B(x,a) = \int_0^1 (1-t)\, d_x^2 f(tx,a)\, dt,$$

is a symmetric matrix depending C^∞ on x and a, $B(0,a) = Q(a)$. Thus we are finished if

$$(1.2.11) \qquad R'Q(a)R = B(x,a).$$

For $x = 0$, $R = I$ is a solution. The differential of the left-hand side with respect to R at $R = I$ is equal to $S \to S'Q(a) + Q(a)S$. This mapping is surjective from the space of all matrices to the space of symmetric matrices, because $S'Q + QS = C$ has the solution $S = \tfrac{1}{2}Q^{-1}C$. Application of the implicit function theorem therefore gives a solution Q of (1.2.11) depending C^∞ on x and a, for x in a neighborhood of $x = 0$. $\qquad\square$

So application of Lemma 1.2.2 to condition (1.2.8) gives for each $a_0 \in A$ a neighborhood A_0 of a_0, finitely many open subsets $X^{(j)}$, $j = 1, \ldots, k$ of $\mathbb{R}^n$, and corresponding C^∞ functions $x^{(j)} : A_0 \to X^{(j)}$, $y^{(j)} : X^{(j)} \times A_0 \to \mathbb{R}^n$, such that

$$(1.2.12) \qquad d_x f(x, a) \neq 0 \text{ for } a \in A_0, \qquad x \in K \setminus \bigcup_{j=1}^{k} X^{(j)},$$

and

$$(1.2.13) \quad \begin{array}{l} x \to y^{(j)}(x, a) \text{ is a diffeomorphism from } X^{(j)} \text{ onto} \\[4pt] \text{an open subset of } \mathbb{R}^n \text{ such that } y^{(j)}(x^{(j)}(a), a) = 0, \\[4pt] d_x y^{(j)}(x^{(j)}(a), a) = I \text{ and} \\[4pt] f(x, a) = f(x^{(j)}(a), a) + \tfrac{1}{2} \langle Q^{(j)}(a) y, y \rangle, \text{ with} \\[4pt] y = y^{(j)}(x, a), \ Q^{(j)}(a) = d_x^2 f(x^{(j)}(a), a), \text{ for all } a \in A_0. \end{array}$$

Take a partition of unity on X consisting of functions $\varphi^{(0)}, \varphi^{(1)}, \ldots, \varphi^{(k)}$ such that

$$(1.2.14) \qquad \operatorname{supp} \varphi^{(j)} \subset X^{(j)} \text{ for } j = 1, \ldots, k,$$

$$(1.2.15) \qquad d_x f(x, a) \neq 0 \text{ for } x \in K \cap \operatorname{supp} \varphi^{(0)}, \quad a \in A_0.$$

Then

$$(1.2.16) \qquad I(a, t) = \sum_{j=0}^{k} I^{(j)}(a, t)$$

such that

$$(1.2.17) \qquad I^{(0)}(a, t) = \int e^{it f(x, a)} \varphi^{(0)}(x) \cdot g(x, a, t) \, dx = O(t^{-N})$$

for $t \to \infty$, uniformly in $a \in A_0$, and for all N;

$$(1.2.18) \qquad \begin{array}{c} I^{(j)}(a, t) = e^{it f(x^{(j)}(a), a)} \cdot \int e^{it \langle Q^{(j)}(a) y, y \rangle / 2} g^{(j)}(y, a, t) \, dy \\[6pt] \text{for } j = 1, \ldots, k, \end{array}$$

where the amplitude $g^{(j)}$ is given by

$$(1.2.19) \qquad g^{(j)}(y^{(j)}(x, a), a, t) \cdot |\det d_x y^{(j)}(x, a)| = \varphi^{(j)}(x) \cdot g(x, a, t).$$

Therefore we have reduced the study of (1.2.1) to the case that $f(x,a) = \frac{1}{2}\langle Q(a)x, x\rangle$.

Lemma 1.2.3. *If A is a nonsingular symmetric $n \times n$ matrix then the Fourier transform of*

$$(1.2.20) \qquad\qquad x \to e^{i\langle Ax,x\rangle/2}$$

is equal to

$$(1.2.21) \qquad \xi \to (2\pi)^{n/2} \cdot |\det A|^{-1/2} \cdot e^{\frac{\pi i}{4} \operatorname{sgn} A} e^{-i\langle A^{-1}\xi,\xi\rangle/2}.$$

Here $\operatorname{sgn} A$ is the number of positive eigenvalues of A minus the number of negative eigenvalues of A.

Proof. On some orthonormal base A has diagonal form, so $\langle Ax, x\rangle = \sum_{j=1}^{n} a_j x_j^2$ and the Fourier transform of (1.2.20) is the product of the 1-dimensional Fourier transforms of the functions $x_j \to e^{ia_j x_j^2/2}$.

Now the 1-dimensional Fourier transform of $x \to e^{-zx^2/2}$ is complex analytic in z in $\operatorname{Re} z > 0$ and equal to

$$(1.2.22) \qquad\qquad \xi \to \left(\frac{2\pi}{z}\right)^{1/2} e^{-\xi^2/2z}$$

for real positive z, so it is equal to (1.2.22) for $\operatorname{Re} z > 0$. Furthermore, as a temperate distribution, it depends continuously on z in $\operatorname{Re} z \geq 0$ so we conclude that if $\operatorname{Re} z = 0$, $z \neq 0$ then the Fourier transform is still equal to (1.2.22), which actually is an analytic function of ξ. For the complex analytic continuation $z \to z^{-1/2}$ to $\operatorname{Re} z \geq 0$, $z \neq 0$ we have $(-ia)^{-1/2} = |a|^{-1/2}e^{\frac{\pi i}{4}\operatorname{sgn} a}$ if a is real. This completes the proof of the lemma.

Proposition 1.2.4. *Suppose g satisfies (1.2.2) and (1.2.3) with $\delta < 1/2$. Let Q be nonsingular and symmetric, depending continuously on a. Then*

$$\int e^{it\langle Q(a)x,x\rangle/2} g(x,a,t) \sim \left(\frac{2\pi}{t}\right)^{n/2} \cdot |\det Q(a)|^{-1/2}$$

$$(1.2.23)$$

$$\cdot e^{\frac{\pi i}{4}\operatorname{sgn} Q(a)} \cdot \sum_{k=0}^{\infty} \frac{1}{k!}(R^k g)(0,a,t) \cdot t^{-k}$$

for $t \to \infty$, *uniformly in* $a \in A$. *Here*

$$R = i\left\langle Q(a)^{-1}\frac{\partial}{\partial x}, \frac{\partial}{\partial x}\right\rangle/2$$

which is a second-order partial differential operator in x.

Proof.

$$\int e^{itQ/2} \cdot g \, dx = \int \mathcal{F}(e^{itQ/2})(\xi) \cdot (\mathcal{F}^{-1}g)(\xi, a, t) \, d\xi.$$

Here $\mathcal{F}$ denotes Fourier transformation. Apply Lemma 1.2.3 and use the Taylor series

$$e^{-i\langle Q^{-1}\xi, \xi\rangle/2t} = \sum_{k=0}^{\infty} \frac{1}{k!}(-i\langle Q^{-1}\xi, \xi\rangle/2)^k \cdot t^{-k}$$

Then apply partial integration and use that

$$\int (\mathcal{F}^{-1}f)(\xi) \, d\xi = f(0)$$

to complete the proof.

Historical remark. The method of stationary phase goes back to Stokes and Kelvin, with refinements by van der Corput. See Erdelyi [28], Section 2.9 for a short review.

1.3. The wave front set of a distribution

Let $u \in \mathcal{D}'(X)$, X open in $\mathbb{R}^n$. According to the Paley–Wiener theorem u is C^∞ in a neighborhood of x, in other words, $x \notin \text{sing supp}\, u$, if and only if there exists $\varphi \in C_0^\infty(X)$ such that $\varphi(x) \neq 0$ and

(1.3.1) $\qquad \mathcal{F}(\varphi u)(\xi) = O(|\xi|^{-N})$ for $|\xi| \to \infty$, all N.

(In fact (1.3.1) means that $\varphi u \in C_0^\infty(\mathbb{R}^n)$. An equivalent formulation is: there is a neighborhood U of x such that (1.3.1) holds for every $\varphi \in C_0^\infty(U)$.)

Now (1.3.1) is equivalent to

(1.3.2) $\qquad \mathcal{F}(\varphi u)(\tau\xi) = \langle e^{-i\tau\langle \cdot, \xi\rangle}\varphi, u\rangle = O(\tau^{-N})$

for $\tau \to \infty$, uniformly in $|\xi| = 1$, for all N. So we tested the distribution u with the oscillatory test function $e^{-i\tau\langle x, \xi\rangle}\varphi(x)$ and then investigated the asymptotic behavior letting the frequency variable τ go to ∞. The function

φ was used to obtain a *localization with respect to the x-variables*. Notice that $\langle x, \xi \rangle = $ constant are the fronts of constant phase of the oscillating test function, which are orthogonal to the vector ξ. Now it turns out that it is very fruitful not only to localize with respect to x but also with respect to the direction from which the testing waves pass over the distribution u. This leads to the following definition.

Definition 1.3.1. If $u \in \mathcal{D}'(X)$, then the *wave front set* $WF(u)$ of u is defined as the complement in $X \times (\mathbb{R}^n \setminus \{0\})$ of the collection of all $(x_0, \xi_0) \in X \times (\mathbb{R}^n \setminus \{0\})$ such that for some neighborhood U of x_0, V of ξ_0 we have for each $\varphi \in C_0^\infty(U)$ and each N:

$$(1.3.3) \qquad \mathcal{F}(\varphi u)(\tau \xi) = O(\tau^{-N}) \text{ for } \tau \to \infty, \text{ uniformly in } \xi \in V.$$

Proposition 1.3.1. $WF(u)$ *is a closed cone in* $X \times (\mathbb{R}^n \setminus \{0\})$,

$$(1.3.4) \qquad\qquad \text{sing supp } u = \pi(WF(u)),$$

and finally

$$(1.3.5) \quad WF(u|_Y) = WF(u) \cap \pi^{-1}(Y) \text{ for every open subset } Y \text{ of } X.$$

Here π is the projection $(x, \xi) \to x$ from $X \times (\mathbb{R}^n \setminus \{0\})$ onto its first factor. Also, $u|_Y$ is short for $u|_{C_0^\infty(Y)}$.

A subset Γ of $X \times (\mathbb{R}^n \setminus \{0\})$ is called a cone if

$$(1.3.6) \qquad\qquad (x, \xi) \in \Gamma \Rightarrow (x, \tau\xi) \in \Gamma \text{ for all } \tau > 0.$$

The proof of Proposition 1.3.1 is immediate. To obtain a coordinate invariant definition of wave front sets of distributions of manifolds, we give the following variant of (1.3.3).

Proposition 1.3.2. $(x_0, \xi_0) \notin WF(u)$ *if and only if for any real-valued* C^∞ *function* $\psi(x, 0)$ *of* $(x, a) \in \mathbb{R}^n \times \mathbb{R}^p$ *with* $d_x \psi(x_0, a_0) = \xi_0$ *there is an open neighborhood* U_0 *of* x_0, A_0 *of* a_0 *such that for any* $\varphi \in C_0^\infty(U_0)$ *we have*

$$(1.3.7) \qquad\qquad \langle e^{-i\tau\psi(\cdot, a)}\varphi, u \rangle = O(\tau^{-N}) \text{ for } \tau \to \infty,$$

uniformly in $a \in A_0$.

Proof. This is a direct application of Proposition 2.1.1. Suppose $(x_0, \xi_0) \notin WF(u)$ (the "if" part is trivial). Let $\varphi' \in C_0^\infty(U_0)$ be equal to 1 on a neighborhood of supp φ. Then we get

$$\langle e^{-i\tau\psi(\cdot, a)}\varphi, u \rangle = \langle \mathcal{F}^{-1}(e^{-i\tau\psi}\varphi'), \mathcal{F}(\varphi u) \rangle$$

$$= (2\pi)^{-n} \iint e^{i\langle x, \xi \rangle} e^{-i\tau\psi(x, a)} \varphi'(x) \, \mathcal{F}(\varphi u)(\xi) \, dx \, d\xi$$

$$= (2\pi)^{-n} \tau^n \iint e^{i\tau[\langle x, \xi \rangle - \psi(x, a)]} \varphi'(x) \cdot \mathcal{F}(\varphi u)(\tau\xi) \, dx \, d\xi,$$

using the substitution $\xi \to \tau\xi$.

Now we apply Proposition 2.1.1, or rather its proof, to the integral

$$(1.3.8) \qquad I(\tau, \xi, a) = \int e^{i\tau[\langle x, \xi \rangle - \psi(x, a)]} \varphi'(x) \, dx.$$

Choosing U_0, A_0 small enough we can obtain in view of $d_x\psi(x_0, a_0) = \xi_0$ that $|d_x\psi(x, a) - \xi| \geq \varepsilon$ if $(x, a) \in U_0 \times A_0$, $\xi \notin V$. Here V is the neighborhood of ξ_0 as in Definition 1.3.2. So if $\xi \notin V$, $a \in A_0$ then application of the partial integrations in the proof of Proposition 2.1.1 with $L = |\xi - d_x\psi|^{-2}\langle \xi - d_x\psi, \frac{\partial}{\partial x} \rangle$ leads to an estimate of the form

$$(1.3.9) \qquad |I(\tau, \xi, a)| \leq C_k \cdot \tau^{-k}(1 + |\xi|)^{-k}, \quad a \in A_0, \ \xi \notin V, \ \tau \geq 1.$$

On the other hand $|\mathcal{F}(\varphi u)(\tau\xi)| \leq C(1 + |\tau\xi|)^\ell$, some ℓ, and

$$(1.3.10) \qquad |\mathcal{F}(\varphi u)(\tau\xi)| \leq C_k' \cdot \tau^{-k}, \qquad \xi \in V, \ \tau \geq 1,$$

any k if $U_0 \subset U$, U as in Definition 1.3.1. These two estimates together immediately yield (1.3.7). $\qquad \square$

So if X is a manifold, $u \in \mathcal{D}'(X, \Omega_\alpha)$, then we can take Proposition 1.3.2 as the definition of $WF(u)$, the only difference is that we should take $\varphi \in C_0^\infty(U_0, \Omega_{1-\alpha})$. This definition is coordinate-invariant by its very formulation and agrees with Definition 1.3.1 for distributions in $\mathbb{R}^n$. We see that $WF(u)$ is a closed conic subset of $T^*(X) \setminus 0$ = the *cotangent bundle* $T^*(X)$ of X minus the zero section. Also (1.3.4) and (1.3.5) hold with π equal to the bundle projection: $T^*(X) \to X$.

The cone axes $\alpha(x, \xi) = \{(x, \tau\xi); \ \tau > 0\}$ through points of $T^*(X) \setminus 0$ together form a bundle $S^*(X)$ over X, with projection $\beta: \alpha(x, \xi) \to x$ and fiber isomorphic to the sphere in $T_x(X)^*$ with radius 1. So the bundle mapping π factors through the bundle mappings α and β ($\pi = \beta \circ \alpha$). $\beta \colon S^*(X) \to X$ has the useful property that

$$T^*(X) \setminus 0$$
$$\downarrow \ \alpha, \ \text{fiber} = R^+$$
$$S^*(X)$$
$$\downarrow \ \beta, \ \text{fiber} = S^{n-1}$$
$$X$$

its fibers are compact. Because β is a fibration, this implies that β is a *proper mapping*, that is, preimages of compact subsets of X are compact in $S^*(X)$.

The concept of wave front sets can be used to define a sheaf $\mathcal{S}$ on $S^*(X)$ which is analogous to the sheaf $\mathcal{C}$ of Sato [71]–[73] in the category of hyperfunctions. Let U be an open subset of $S^*(X)$, which can also be regarded as a conic open subset of $T^*(X) \setminus 0$. Call two distributions u_1, u_2 on $\mathcal{D}'(X)$ *equivalent over U*, notation $u_1 \equiv u_2$ in U, if $WF(u_1 - u_2) \cap U = \emptyset$. The equivalence classes with respect to this equivalence relation form a space $\mathcal{S}(U)$ and we have a natural mapping $\rho_{U,U'} \colon \mathcal{S}(U) \to \mathcal{S}(U')$ if $U' \subset U$. One can prove that the $\mathcal{S}(U)$ together with the "restriction mappings" $\rho_{U,U'}$ form a presheaf and hence define a sheaf over $S^*(X)$. Sections of this sheaf $\mathcal{S}$ over the whole of $S^*(X)$ are naturally identified with elements of $\mathcal{D}'(X)/C^\infty(X)$ and the support of such a section is equal to the wave front set of the corresponding distribution.

Historically, Sato first defined his sheaf $\mathcal{C}$ over $S^*(X)$, X a real analytic manifold. Global sections of $\mathcal{C}$ correspond with hyperfunctions on X modulo real-analytic functions. For the supports of global sections he derived the properties analogous to Theorems 1.3.4 and 1.4.1 below. This inspired Hörmander to his definition of wave front sets (and the sheaf $\mathcal{S}$) in [40], Section 2.5. A corresponding definition in Gevrey classes (including the real-analytic case) was given in [43].

We now follow the presentation of the calculus of wave front sets of Gabor [31].

Definition 1.3.2. Let Γ be a closed cone in $T^*(X) \setminus 0$. Define $\mathcal{D}'_\Gamma(X) = \{u \in \mathcal{D}'(X); WF(u) \subset \Gamma\}$. In $\mathcal{D}'_\Gamma(X)$ we take the topology defined by the seminorms of the weak topology in $\mathcal{D}'(X)$ together with the seminorms

$$(1.3.11) \qquad u \mapsto \sup_{\tau \geq 1, a \in A} \tau^N \left| \langle e^{-i\tau\psi(\cdot,a)} \varphi, u \rangle \right|,$$

where $\varphi \in C_0^\infty(X)$, A compact in $\mathbb{R}^p$, $(x, d_x\psi(x,a)) \notin \Gamma$ for $(x,a) \in$ supp $\varphi \times A$.

In other words, we add as seminorms the smallest possible constants in (1.3.7). An equivalent definition can be given taking the best possible constants in (1.3.3), by taking supp φ contained in a coordinate neighborhood and $\mathcal{F}(\varphi \cdot u) =$ Fourier transform of $\varphi \cdot u$ in the corresponding local coordinates.

Proposition 1.3.3. *Let X, Y be C^∞ manifolds, Φ a C^∞ mapping: $X \to Y$, denote*

(1.3.12)
$$N = \{(y, \eta) \in T^*(Y) \setminus 0; \quad y = \Phi(x),$$
$${}^t D\Phi_x\eta = 0 \quad \text{for some} \ x \in X\}.$$

Let Γ be a closed cone in $T^*(Y) \setminus 0$ such that $\Gamma \cap N = \emptyset$. Then the pullback $\Phi^*: C^\infty(Y) \to C^\infty(X)$ has a unique continuous extension: $\mathcal{D}'_\Gamma(Y) \to \mathcal{D}'(X)$, and supp$(\Phi^*v) \subset \Phi^{-1}(\text{supp}\,v)$ for each $v \in \mathcal{D}'_\Gamma(Y)$. If

(1.3.13) $\quad \tilde{\Gamma} = \{(x, \xi) \in T^*(X) \setminus 0; \quad \exists \eta : {}^t D\Phi_x\eta = \xi, \ (\Phi(x), \eta) \in \Gamma\}$

then Φ^* is in fact continuous: $\mathcal{D}'_\Gamma(Y) \to \mathcal{D}'_{\tilde{\Gamma}}(X)$.

Proof. Using a partition of unity in Y it suffices to prove it for v such that supp v is compact and contained in a coordinate neighborhood, and there we define Φ^*v formally by

(1.3.14)
$$\langle \Phi^*v, \varphi \rangle = (2\pi)^{-n} \iint e^{i\langle \Phi(x), \eta \rangle} \varphi(x)$$
$$\cdot (\mathcal{F}v)(\eta) \, d\eta \, dx, \quad \varphi \in C_0^\infty(X).$$

Here $n = \dim Y$. It is obvious that we only need to consider φ with small support near points in $\Phi^{-1}(\text{supp}\,v)$. Making the partition of unity in Y fine enough we can therefore reduce to the situation that ${}^t D\Phi_x\eta \neq 0$ for (x, η) in the closure of the set

$$\{(x, \eta); \ x \in \text{supp}\,\varphi, \ |\eta| = 1, \ (y, \eta) \notin WF(v) \ \text{for some} \ y \in Y\}.$$

(Here it is used that Γ is closed and $\Gamma \cap N = \emptyset$.) It follows that there exists $\chi \in C^\infty(\mathbb{R}^n)$ such that $\chi = 1$ on a neighborhood of 0 in $\mathbb{R}^n$, $\chi(\eta)$ is homogeneous of degree 0 for large $|\eta|$,

$$|\chi(\eta) \cdot (\mathcal{F}v)(\eta)| \le c_N (1 + |\eta|)^{-N} \quad \text{for any} \ N$$

and finally

$$|{}^t D\Phi_x \eta| \geq C|\eta| \quad \text{for} \quad x \in \operatorname{supp}\varphi, \ \eta \in \operatorname{supp}(1-\chi).$$

Let $L = \Sigma L_j(x,\eta)\partial/\partial x_j$ be such that $L_j \in C^\infty(X \times \mathbb{R}^n \setminus \{0\})$, L_j is homogeneous in η of degree -1 and finally $L\cdot\langle\Phi(x),\eta\rangle = 1/i$ on $\operatorname{supp}(1-\chi)$. Then

(1.3.15)
$$\langle\Phi^* v, \varphi\rangle = \iint e^{i\langle\Phi(x),\eta\rangle}\varphi \cdot \chi \cdot \mathcal{F}v \, d\eta \, dx$$
$$+ \iint e^{i\langle\Phi(x),\eta\rangle}({}^tL)^\rho\varphi \cdot (1-\chi) \cdot \mathcal{F}v \, d\eta \, dx$$

is absolutely convergent for sufficiently large ρ and defines the desired continuous extension Φ^*. (Note that (1.3.15) is equal to (1.3.14) when $v \in C_0^\infty(\mathbb{R}^n)$.) The uniqueness of the extension follows from the remark that $v * \psi_\varepsilon \to v$ in $\mathcal{D}'_\Gamma(\mathbb{R}^n)$ if $v \in \mathcal{D}'_\Gamma(\mathbb{R}^n)$, $\psi_\varepsilon(y) = \varepsilon^{-n}\psi(y/\varepsilon)$, $\psi \in C_0^\infty(\mathbb{R}^n)$, $\int \psi \, dy = 1$, so $C^\infty(\mathbb{R}^n)$ is dense in $\mathcal{D}'_\Gamma(\mathbb{R}^n)$. $\qquad\Box$

Proposition 1.3.4. *The push-forward* $\Phi_* = {}^t(\Phi^*)$ *is a continuous mapping from the space of* $u \in \mathcal{D}'(X)$ *such that* $\Phi: \operatorname{supp} u \to Y$ *is a proper mapping into* $\mathcal{D}'(Y)$. *For such* u *we have*

(1.3.16)
$$WF(\Phi_* u) \subset \{(y,\eta) \in T^*(Y)\setminus 0; \ y = \Phi(x)$$
$$\text{and} \ (x, {}^t D\Phi_x \eta) \in WF(u) \ \text{for some} \ x \in X\}.$$

Proof.
$$\langle\Phi_* u, \varphi(y)e^{-i\tau\psi(y)}\rangle = \langle u, \varphi(\Phi(x))e^{-i\tau\psi(\Phi(x))}\rangle.$$
Note that $d_x\psi(\Phi(x)) = d\psi_{\Phi(x)} \circ D\Phi_x = {}^t D\Phi_x \cdot d\psi_{\Phi(x)}$. $\qquad\Box$

Proposition 1.3.5. *If* $u \in \mathcal{D}'(X)$, $v \in \mathcal{D}'(Y)$ *then*

$$WF(u \otimes v) \subset (WF(u) \times WF(v)) \cup (WF(u) \times \operatorname{supp}_0 v) \cup (\operatorname{supp}_0 u \times WF(v)).$$

Here $\operatorname{supp}_0 u = \{(x,0) \in T^*(X); \ x \in \operatorname{supp} u\}$ *and analogously* $\operatorname{supp}_0 v = \{(y,0) \in T^*(Y); \ y \in \operatorname{supp} v\}$.

The proof is immediate. From Propositions 1.3.3 and 1.3.5 we now obtain:

Theorem 1.3.6. *Let* Γ_1, Γ_2 *be closed cones in* $T^*(X)\setminus 0$ *such that*

$$\Gamma_1 + \Gamma_2 := \{(x,\xi_1+\xi_2); \ (x,\xi_1) \in \Gamma_1, \ (x,\xi_2) \in \Gamma_2\}$$

does not meet the zero section in $T^*(X)$. *Then there is a unique continuous mapping:*

$$\mathcal{D}'_{\Gamma_1}(X) \times \mathcal{D}'_{\Gamma_2}(X) \to \mathcal{D}'(X)$$

extending the product $(u_1, u_2) \mapsto u_1 \cdot u_2 \colon C^\infty(X) \times C^\infty(X) \to C^\infty(X)$. *Moreover* $(\Gamma_1 + \Gamma_2) \cup \Gamma_1 \cup \Gamma_2$ *is a closed cone in* $T^*(X) \setminus 0$ *and the product is in fact continuous:*

$$\mathcal{D}'_{\Gamma_1}(X) \times \mathcal{D}'_{\Gamma_2}(X) \to \mathcal{D}'_{(\Gamma_1 + \Gamma_2) \cup \Gamma_1 \cup \Gamma_2}(X).$$

Proof. $u_1 \cdot u_2 = \Delta^*(u_1 \otimes u_2)$ where $\Delta \colon X \to X \times X \colon x \mapsto (x, x)$ is the diagonal map. That $(\Gamma_1 + \Gamma_2) \cup \Gamma_1 \cup \Gamma_2$ is closed follows from $\overline{\Gamma_1 + \Gamma_2} \subset (\Gamma_1 + \Gamma_2) \cup \Gamma_1 \cup \Gamma_2$, which we now prove. Suppose $\xi_1^j + \xi_2^j \to \xi \neq 0$ for a sequence $\xi_1^j \in \Gamma_1$, $\xi_2^j \in \Gamma_2$. Suppose $|\xi_1^j| \to \infty$ for a subsequence; after division by $|\xi_1^j|$ and passing again to a subsequence if necessary, this leads to $0 \in \Gamma_1 + \Gamma_2$, a contradiction. If $|\xi_1^j| \to 0$ for a subsequence then $\xi \in \Gamma_2$. If finally $\xi_1^j \to \xi_1 \neq 0$, $\xi_2^j \to \xi_2 \neq 0$ for a subsequence then $\xi \in \Gamma_1 + \Gamma_2$. $\square$

In the following theorems concerning continuous mappings

$$A \colon C_0^\infty(Y) \to \mathcal{D}'(X) \text{ with distribution kernel } K_A \in \mathcal{D}'(X \times Y)$$

it is convenient to introduce the following notations.

(1.3.17)
$$WF'(A) = \{((x, \xi), (y, \eta)) \in (T^*(X) \times T^*(Y)) \setminus 0;$$
$$(x, y; \xi, -\eta) \in WF(K_A)\},$$

(1.3.18) $\quad WF'_X(A) = \{(x, \xi) \in T^*(X) \setminus 0; \ \exists y \in Y \colon (x, y; \xi, 0) \in WF(K_A)\}$

(1.3.19) $\quad WF'_Y(A) = \{(y, \eta) \in T^*(Y) \setminus 0; \ \exists x \in X \colon (x, y; 0, \eta) \in WF(K_A)\}.$

If $R_1 \subset U \times V$, $R_2 \subset V \times W$ are relations then the *composition* $R_1 \circ R_2 \subset U \times W$ is defined by

(1.3.20) $\quad R_1 \circ R_2 = \{(u, w) \in U \times W; \ \exists v \in V \colon (u, v) \in R_1 \text{ and } (v, w) \in R_2\}.$

Theorem 1.3.7. *Let* X, Y, Z *be* C^∞ *manifolds,* A *and* B *continuous linear mappings:* $C_0^\infty(Y) \to \mathcal{D}'(X)$ *and* $C_0^\infty(Z) \to \mathcal{D}'(Y)$, *respectively. If* $WF'_Y(A) \cap WF'_Y(B) = \emptyset$ *and the projection into* $X \times Z$ *from the diagonal is* $X \times Y \times Y \times Z$ *is a proper mapping, then* $A \circ B$ *is a well-defined* *continuous linear mapping:* $C_0^\infty(Z) \to \mathcal{D}'(X)$. *Moreover* $\operatorname{supp} K_{A \circ B} \subset$ $\operatorname{supp} K_A \circ \operatorname{supp} K_B$ *and*

$$
(1.3.21) \qquad
\begin{aligned}
WF'(A \circ B) \subset WF'(A) \circ WF'(B) \cup (WF'_X(A) \times O_{T^*(Z)}) \\
\cup (O_{T^*(X)} \times WF'_Z(B)).
\end{aligned}
$$

Proof. $K_{A \circ B} = \pi_* \Delta^* (K_A \otimes K_B)$, where $\Delta \colon (x,y,z) \mapsto (x,y,y,z) \colon X \times Y \times Z \to X \times Y \times Y \times Z$ and

$$
\pi \colon (x,y,z) \mapsto (x,z) \colon X \times Y \times Z \to X \times Z. \qquad \square
$$

Corollary 1.3.8. *Let A be a continuous linear mapping:* $C_0^\infty(Y) \to \mathcal{D}'(X)$ *and let Γ be a closed cone in $T^*(Y) \setminus 0$ that does not meet $WF'_Y(A)$. Then A can be extended to a sequentially continuous mapping:* $\mathcal{D}'_\Gamma(Y) \cap \mathcal{E}'(Y) \to \mathcal{D}'(X)$ *and*

$$
(1.3.22) \qquad WF(Au) \subset (WF'(A) \circ WF(u)) \cup WF'_X(A),
$$

for all $u \in \mathcal{D}'_\Gamma(Y) \cap \mathcal{E}'(Y)$. If in addition the projection from $\operatorname{supp} K_A$ into X is a proper mapping then A can be extended to a sequentially continuous mapping: $\mathcal{D}'_\Gamma(Y) \to \mathcal{D}'(X)$ *and (1.3.22) holds for all $u \in \mathcal{D}'_\Gamma(Y)$.*

Proof. Apply Theorem 1.3.7 with $Z = \{point\}$. $\qquad \square$

Remark. In view of its role in Theorem 1.3.7 and Corollary 1.3.8, $WF'(A)$ will be called the *wave front relation* of the operator A, it describes how A propagates wave front sets of distributions on which it acts (ignoring $WF'_X(A)$ for the moment).

Chapter 2

Local Theory of Fourier Integrals

2.1. Symbols

In this section we generalize the classes of amplitude functions encountered in the Introduction and in Section 1.2, and we collect some useful properties of these "symbol spaces."

Definition 2.1.1. A *conic manifold* is a C^∞ paracompact manifold V together with a proper and free C^∞ action of $\mathbb{R}_+$ on V. ($\mathbb{R}_+$ is regarded as a multiplicative group.) It follows that the orbit space $V' = V/\mathbb{R}_+$ has a C^∞ structure making V into a fiber bundle over V' with $\mathbb{R}_+$ as fiber, and the mapping α assigning to each $v \in V$ the orbit through v as projection. The orbit $\alpha(v)$ is also called the *cone axis* through v.

Piecing together local sections by means of a partition of unity in V' one can always construct a global section $V' \to V$, making the bundle trivial. For this, we observe that if s_α, s_β are local sections, then s_α/s_β is a strictly positive function. So, if φ_α is the partition of unity, then

$$s = \prod_\alpha (s_\alpha/s_\beta)^{\varphi_\alpha} \cdot s_\beta$$

is independent of β and defines the desired global section s. However, we avoid writing $V = V' \times \mathbb{R}_+$ because there may be no preferred choice of the unit section.

A function f on V is called *homogeneous of degree* μ if $\tau^* f = t^\mu f$ for all $\tau \in \mathbb{R}_+$. Here the pullback $\tau^* f$ of f by means of τ is defined by

$$(2.1.1) \qquad (\tau^* f)(v) = f(\tau v), \qquad v \in V.$$

If L is a smooth vector field on V and f is a smooth function on V, then $\tau^*(Lf) = (\tau^* L)(\tau^* f)$, where

$$(2.1.2) \qquad (\tau^* L)(v) = D\tau_v^{-1}(L(\tau v)), \qquad v \in V.$$

P. Buser, *Fourier Integral Operators*, Modern Birkhäuser Classics,
DOI 10.1007/978-0-8176-8108-1_3, © Springer Science+Business Media, LLC 2011

Here τ is regarded as a diffeomorphism: $V \to V$, so $D\tau_v$ is a linear mapping: $T_v(V) \to T_{\tau v}(V)$. A vector field L on V will be called homogeneous of degree ν if $\tau^* L = \tau^\nu \cdot L$ for all $\tau \in \mathbb{R}_+$, it follows that Lf is homogeneous of degree $\mu + \nu$ if f is homogeneous of degree μ and L is homogeneous of degree ν. Note that homogeneous vector fields of degree 0 induce flows in V which commute with the action of $\mathbb{R}_+$ on V.

Definition 2.1.2. *Let V be a conic manifold, $\mu, \rho \in \mathbb{R}$, $0 \le \rho \le 1$. A symbol on V of order μ and type ρ is a function $a \in C^\infty(V)$ such that*

$$(2.1.3) \qquad \tau^*(L_k \cdot \ldots \cdot L_1 a) = O(\tau^{\mu - k\rho}) \quad \text{for} \quad \tau \to \infty,$$

locally uniformly in V and for all C^∞ vector fields $L_1, \ldots, L_k$ in V that are homogeneous of degree -1. The space of these symbols is denoted by $S_\rho^\mu(V)$.

Note that $a \in S_1^\mu(V)$ if a is homogeneous of degree μ. In most applications we will only use symbols of type 1 and we will write $S_1^\mu(V) = S^\mu(V)$. Note also that $a \in S_\rho^\mu(V)$ if and only if

$$(2.1.4) \qquad \tau^*(L_k \cdot \ldots \cdot L_1 a) = O(\tau^{\mu + k\delta}) \quad \text{for} \quad \tau \to \infty,$$

locally uniformly in V and for all homogeneous C^∞ vector fields L_j of degree 0, $\delta = 1 - \rho$. From an a priori point of view vector fields of degree 0 are more natural, but we have chosen $\rho = 1 - \delta$ as type number rather than δ because we shall make extensive use of homogeneous vector fields of degree -1 later on. The following assertions follow immediately from the definition.

Proposition 2.1.1. *$S_\rho^\mu(V)$ is a linear space. $S_\rho^\mu(V) \subset S_{\rho'}^{\mu'}(V)$ if $\mu \le \mu'$, $\rho \ge \rho'$. If $a \in S_\rho^\mu(V)$ and L is a homogeneous vector field of degree ν, then $La \in S_\rho^{\mu + \nu + (1-\rho)}(V)$. If $a \in S_\rho^\mu(V)$, $b \in S_\rho^{\mu'}(V)$, then $a \cdot b \in S_\rho^{\mu + \mu'}(V)$. Finally, if W is another conic manifold and χ is a C^∞ mapping: $V \to W$ commuting with the actions of $\mathbb{R}_+$ on V, respectively, W, then $\chi^*: a \to a \circ \chi$ maps $S_\rho^\mu(W)$ into $S_\rho^\mu(V)$.*

The following propositions are the analogues of Theorems 2.7 and 2.9 in Hörmander [41]. We denote $S^{-\infty}(V) = \bigcap_{\mu \in \mathbb{R}} S_\rho^\mu(V)$, which does not depend on ρ.

Proposition 2.1.2. *Suppose* $a_j \in S_\rho^{\mu_j}(V)$, $j = 0, 1, 2, \ldots$, *and* $\mu_j \searrow -\infty$ *for* $j \to \infty$. *Then there exists* $a \in S_\rho^{\mu_0}(V)$ *such that*

$$(2.1.5) \qquad a - \sum_{j<k} a_j \in S_\rho^{\mu_k}(V) \quad \text{for all } k = 1, 2, \ldots.$$

In this case we say $a \sim \sum a_j$.

Proof. We may write $V = V' \times \mathbb{R}_+$. Let K_j be an increasing sequence of compact subsets of V' such that every compact subset of V' is contained in one of them. Let $\mathcal{L}_j$ be a finite set of C^∞ vector fields on V' such that the $L(x)$, $L \in \mathcal{L}_j$ span $T_x(V')$ for every $x \in K_j$. Choose $\varphi \in C^\infty(\mathbb{R}_+)$ equal to 0 when $\tau \leq 1/2$ and equal to 1 when $\tau \geq 1$. We can then select a sequence $\tau_j \to \infty$ increasing so rapidly that

$$(2.1.6) \qquad \left| L_\ell \cdots L_1 \left(\frac{\partial}{\partial \tau} \right)^k \varphi(\tau_j^{-1}\tau) a_j(x, \tau) \right| \leq 2^{-j} \tau^{\mu_j - 1 - \rho k + (1-\rho)\ell}$$

for $x \in K_i$, $L_1, \ldots, L_\ell \in \mathcal{L}_1$, $k + \ell + i \leq j$, $\tau \geq 1$.

In fact, since there is only a finite number of conditions for given j, we only need to use the fact that $\tau^k \frac{\partial^k}{\partial \tau^k} \varphi(\tau_j^{-1}\tau)$ is uniformly bounded for each j. We can now take

$$(2.1.7) \qquad a(x, \tau) = \sum_{j=0}^\infty \varphi(\tau_j^{-1}\tau) \cdot a_j(x, \tau)$$

which is a locally finite sum since $\varphi(\tau_j^{-1}\tau) = 0$ for $\tau \leq \frac{1}{2}\tau_j$. The estimates (2.1.3) and (2.1.5) follow by remarking that any vector field on a neighborhood of K_j can be written as a linear combination on K_j of the $L \in \mathcal{L}_j$, with smooth coefficients. So the estimates follow by induction on the number of applied vector fields from the estimates for the $L \in \mathcal{L}_j$. $\square$

Proposition 2.1.3. *Let* $c \in C^\infty(V)$ *be such that for every set of homogeneous vector fields* $L_1, \ldots, L_k$ *and every compact* $K \subset V$ *there are constants* C, μ *such that*

$$(2.1.8) \qquad |(L_k \cdot \ldots \cdot L_1 c)(\tau v)| \leq C \cdot \tau^\mu, \qquad v \in K, \ \tau \geq 1.$$

Then $c \in S^{-\infty}(V)$ *if for any* $\nu \in \mathbb{R}$:

$$(2.1.9) \qquad c(\tau v) = O(\tau^\nu) \qquad \text{for } \tau \to \infty,$$

locally uniformly in $v \in V$.

We conclude that $b \in S_\rho^{\mu_0}(V)$ and $b \sim \sum a_j$ if the a_j are as in Proposition 2.1.2, b satisfies (2.1.8), and if there exists a sequence $\mu_k \searrow -\infty$ for $k \to \infty$ such that

$$(2.1.10) \qquad \Big(b - \sum_{j<k} a_j\Big)(\tau v) = O(\tau^{\mu_k}) \qquad \text{for } \tau \to \infty$$

locally in $v \in V$, for all k.

Proof. If K, K' are compact sets, K in the interior of K', then we have an a priori estimate of the form

$$(2.1.11) \qquad \|f\|_{1,K} \leq C\sqrt{\|f\|_{0,K'} \cdot \|f\|_{2,K'}}.$$

Here $\| \ \|_{j,K}$ denotes some fixed C^j-norm taken over K. It suffices to prove (2.1.11) locally for C^∞ functions with compact support and there it follows from the following observation of E. Landau [52]:

If f is $C^2 : \mathbb{R} \to \mathbb{R}$, $|f(x)| \leq P$, $|f''(x)| \leq Q$ for all $x \in \mathbb{R}$, then $|f'(x)| \leq \sqrt{2PQ}$ for all $x \in \mathbb{R}$.

Proof.

$$f(x) - f(x - \varepsilon) = \varepsilon f'(x) + \tfrac{1}{2}\varepsilon^2 f''(\xi_1), \qquad x - \varepsilon < \xi_1 < x$$
$$f(x + \varepsilon) - f(x) = \varepsilon f'(x) + \tfrac{1}{2}\varepsilon^2 f''(\xi_2), \qquad x < \xi_2 < x + \varepsilon,$$

so

$$f(x + \varepsilon) - f(x - \varepsilon) = 2\varepsilon f'(x) + \tfrac{1}{2}\varepsilon^2(f''(\xi_1) + f''(\xi_2)).$$

This leads to $|f'(x)| \leq P/\varepsilon + \tfrac{1}{2}\varepsilon Q$, take the minimum of the right-hand side for $\varepsilon > 0$. $\qquad\square$

So because c decreases faster than any power of τ and $L_2 L_1 c$ is bounded by some power of τ as $\tau \to \infty$ for any homogeneous vector fields L_1, L_2, and we see that Lc decreases faster than any power of τ as $\tau \to \infty$ for any homogeneous vector field L. By induction it follows that $L_k \ldots L_1 c$ decreases faster than any power of τ as $\tau \to \infty$ for any homogeneous vector fields $L_1, \ldots, L_k$, which means that $c \in S^{-\infty}(V)$.

For the second assertion, note that Proposition 2.1.2 implies that $a \sim \sum a_j$ for some $a \in S_\rho^{\mu_0}(V)$. Then $c = b - a$ satisfies (2.1.8), (2.1.9) in view of (2.1.10), so $b - a \in S^{-\infty}(V)$. This implies that $b \in S_\rho^{\mu_0}(V)$ and $b \sim \sum a_j$. $\qquad\square$

If E is a smooth N-dimensional vector bundle over a paracompact manifold X, then $E \setminus 0$ is a conic manifold with respect to the multiplications with $\tau \in \mathbb{R}_+$ in the fibers. Introducing inner products in the fibers E_x depending smoothly on $x \in X$ (this can be done locally and then globally by piecing together with a partition of unity in X), we see that the bundle of unit spheres is a smooth section: $(E \setminus 0)/\mathbb{R}_+ \to E \setminus 0$, for this reason $(E \setminus 0)/\mathbb{R}_+$ is called the *sphere bundle SE* of E.

$S_\rho^\mu(E)$ will be defined as the set of $a \in C^\infty(E)$ such that $a|_{E \setminus 0} \in S_\rho^\mu(E \setminus 0)$. If $V = U \times \mathbb{R}^N \setminus \{0\}$, U open in $\mathbb{R}^n$ (the local model for $E \setminus 0$), then a vector field

$$L = \sum_{j=1}^{n} a_j(x, \theta) \frac{\partial}{\partial x_j} + \sum_{k=1}^{N} b_k(x, \theta) \frac{\partial}{\partial \theta_k}$$

on V is homogeneous of degree ν if and only if the a_j and b_k are homogeneous of degree ν and $\nu + 1$, respectively.

It follows that $S_\rho^\mu(U \times \mathbb{R}^N)$ is precisely the space of all $a \in C^\infty(U \times \mathbb{R}^N)$ such that for any compact subset K of U and any multi-indices α, β we have an estimate of the form:

$$(2.1.12) \qquad \left| \left(\frac{\partial}{\partial x} \right)^\beta \left(\frac{\partial}{\partial \theta} \right)^\alpha a(x, \theta) \right| \leq C_{\alpha, \beta, K} (1 + |\theta|)^{\mu - \rho|\alpha| + (1 - \rho)|\beta|}$$

for $x \in K$, $\theta \in \mathbb{R}^N \setminus \{0\}$. This is the original form in which Hörmander introduced the symbol spaces $S_{\rho, \delta}^\mu$, $\delta = 1 - \rho$. The inequalities (2.1.12) are invariant under changes of the local trivialization of E in view of Proposition 2.1.1.

The space $S_\rho^\mu(E)$ will be topologized by taking the best possible constants $C_{\alpha, \beta, K}$ in (2.1.12) as semi-norms. With this topology $S_\rho^\mu(E)$ is a Fréchet space. (A similar topology can be introduced on $S_\rho^\mu(V)$ for an arbitrary conic manifold V but we will not use this in the sequel.) A subset M of $S_\rho^\mu(E)$ is bounded if all seminorms are bounded on M. Because of the theorem of Ascoli, the topology of pointwise convergence and that of $S_\rho^{\mu'}(E)$, $\mu' > \mu$ are identical on bounded subsets of $S_\rho^\mu(E)$. This leads to

Proposition 2.1.4. *Let* $a \in S_\rho^\mu(E)$, $\chi \in C^\infty(E)$, $\chi(x, \theta) = 1$ *for* $\theta = 0$ *and* $\chi(x, \theta) = 0$ *for* $|\theta| \geq 1$. *Define* $a_\varepsilon(x, \theta) = \chi(x, \varepsilon\theta) \cdot a(x, \theta)$. *Then* $a_\varepsilon \in S^{-\infty}(E)$, $a_\varepsilon \to a$ *in* $S_\rho^{\mu'}(E)$ *for* $\varepsilon \to 0$, *any* $\mu' > \mu$.

Proof. The functions $\chi_\varepsilon(x, \theta) = \chi(x, \varepsilon\theta)$, $\varepsilon \in [0, 1]$ form a bounded subset in $S^0(E)$. Because multiplication with a is continuous: $S^0(E) \to S_\rho^\mu(E)$, then a_ε, $\varepsilon \in [0,1]$ form a bounded subset of $S_\rho^\mu(E)$. The proof is completed by remarking that $a_\varepsilon \to a$ pointwise. $\qquad\qquad\qquad\qquad\qquad\square$

Corollary 2.1.5. *Let A be a linear mapping from the space $f \in C^\infty(E)$ that vanish for large $|\theta|$ to a Fréchet space F, which is continuous for the $S_\rho^\mu(E)$-topology for every $\mu \in \mathbb{R}$. Then there is a unique extension of A to $S_\rho^\infty(E) := \bigcup_{\mu\in\mathbb{R}} S_\rho^\mu(E)$ that is continuous: $S_\rho^\mu(E) \to F$ for all $\mu \in \mathbb{R}$.*

The generalization of this section to symbol densities of order α is left to the reader. It should be remarked however that the standard density ω_0 of order α in $\mathbb{R}^N$ is homogeneous of degree αN, in the sense that $\tau^* \omega_0 = \tau^{\alpha N} \cdot \omega_0$ for all $\tau \in \mathbb{R}_+$. The relation

$$(2.1.13) \qquad\qquad a \cdot \omega_0 \leftrightarrow a$$

is an identification between $S_\rho^\mu(\mathbb{R}^n \times \mathbb{R}^N, \Omega_\alpha)$ and $S_\rho^{\mu-\alpha N}(\mathbb{R}^n \times \mathbb{R}^N)$.

2.2. Distributions defined by oscillatory integrals

Let X be open in $\mathbb{R}^n$. The integral

$$(2.2.1) \qquad I_\varphi(au) = \iint e^{i\varphi(x,\theta)} a(x,\theta) u(x)\, dx\, d\theta, \qquad u \in C_0^\infty(X)$$

is absolutely convergent if φ is real, $a \in S_\rho^\mu(X \times \mathbb{R}^N)$, and $\mu + N < 0$. In this case $u \to I_\varphi(au)$ is continuous on $C_0^0(X)$ and therefore defines a distribution A in X of order 0. φ will be called a *phase function* if it is homogeneous of degree 1 and has no critical points as a function of (x,θ). In this case the condition on the order of a can be dropped.

Theorem 2.2.1. *Suppose $\varphi \in C^\infty(X \times \mathbb{R}^N \setminus \{0\})$ is real-valued, homogeneous in θ of degree 1 and $d_{(x,\theta)}\varphi(x,\theta) \neq 0$ for all $(x,\theta) \in X \times \mathbb{R}^N \setminus \{0\}$. Suppose $\rho > 0$. Then the mapping $a \to I_\varphi(au)$, defined for symbols a that vanish for large $|\theta|$, can for every $u \in C_0^\infty(X)$ be extended to $S_\rho^\infty(X \times \mathbb{R}^N)$ such that it is continuous on $S_\rho^\mu(X \times \mathbb{R}^N)$ for every μ. Moreover, for every $a \in S_\rho^\mu(X \times \mathbb{R}^N)$ the linear form $A: u \to I_\varphi(au)$ is a distribution of order k if $\mu - k\rho + N < 0$.*

Proof. Let L be a homogeneous C^∞ vector field of degree -1 on $X \times \mathbb{R}^N \setminus \{0\}$ such that $L\varphi = 1$. This exists locally on the sphere bundle $|\theta| = 1$. With a partition of unity the local L can be glued together to a C^∞ vector field on $|\theta| = 1$ such that $L\varphi = 1$. L extends in a unique way to a homogeneous C^∞ vector field of degree -1 on $X \times \mathbb{R}^N \setminus \{0\}$, and because $L\varphi$ is homogeneous of degree 0 we conclude that $L\varphi = 1$ on $X \times \mathbb{R}^N \setminus \{0\}$.

Now let $\chi \in C^\infty(X \times \mathbb{R}^N)$, $\chi = 0$ for $|\theta| \le 1/2$ and $\chi = 1$ for $|\theta| \ge 1$. Then $M = \frac{1}{i}\chi L + (1 - \chi)$ is a first order differential operator with smooth coefficients on $X \times \mathbb{R}^N$ and $Me^{i\varphi} = e^{i\varphi}$. Moreover its transposed operator ${}^tM$ maps $S_\rho^\mu(X \times \mathbb{R}^N)$ into $S_\rho^{\mu-\rho}(X \times \mathbb{R}^N)$ for all μ. We obtain

$$(2.2.2) \qquad I_\varphi(au) = \iint e^{i\varphi}({}^tM)^k(au)\,dx\,d\theta, \qquad \text{any } k,$$

for any a that vanishes for large $|\theta|$, using repeated partial integrations. So in view of Corollary 2.1.5 the mapping $a \to I_\varphi(au)$ has a continuous extension to $S_\rho^\mu(X \times \mathbb{R}^N)$, which is equal to the absolutely convergent integral (2.2.2) if $m - k\rho + N < 0$. Because at most k derivatives of u appear in (2.2.2) this defines a distribution $u \to I_\varphi(au)$ of order k in X. $\square$

If φ, a depend continuously on a parameter t in $C^\infty(X \times \mathbb{R}^N \setminus \{0\})$ and in $S_\rho^\mu(X \times \mathbb{R}^N)$, respectively, then the corresponding distribution A depends continuously on t in view of (2.2.2). This can also be used to justify differentiations with respect to t under the integral sign. Note that we have continuous dependence of a_t in $S_\rho^{\mu'}(X \times \mathbb{R}^N)$ for all $\mu' > \mu$ as soon as the a_t depend continuously on t pointwise and are bounded in $S_\rho^\mu(X \times \mathbb{R}^N)$. See the remark before Proposition 2.1.4.

Theorem 2.2.2. *Let $a \in S_\rho^m$, $\rho > 0$ and φ be as in Theorem 2.2.1. Then $WF(A)$ is contained in the closed conic subset*

$$\{(x, d_x\varphi(x,\theta)) \in T^*(X) \setminus 0; \quad (x,\theta) \in \text{ess supp } a, \ d_\theta\varphi(x,\theta) = 0\}$$

of $T^*(X) \setminus 0$. Here ess supp a is defined as the smallest conic subset of $X \times \mathbb{R}^N \setminus \{0\}$ outside of which a is of class $S^{-\infty}$.

Proof. We have to prove that the integral

$$(2.2.3) \quad \begin{aligned} \langle e^{-i\tau\psi}u, A \rangle &= \iint e^{i[\varphi(x,\theta)-\tau\psi(x,\sigma)]}a(x,\theta)u(x)\,dx\,d\theta \\ &= \tau^N \iint e^{i\tau[\varphi(x,\theta)-\psi(x,\sigma)]}a(x,\tau\theta)u(x)\,dx\,d\theta. \end{aligned}$$

is rapidly decreasing as $\tau \to \infty$, uniformly for σ in a neighborhood of σ_0, if supp u is contained in a sufficiently small neighborhood U of x_0 and

$$d_x\psi(x_0,\sigma_0) \ne d_x\varphi(x_0,\theta) \text{ for any } \theta \text{ such that}$$

$$(x_0,\theta) \in \text{ess supp } a, d_\theta\varphi(x_0,\theta) = 0.$$

But this means that $d_{(x,\theta)}\chi(x_0,\theta,\sigma_0) \neq 0$ for all θ with $(x_0,\theta) \in$ ess supp a, if we write

$$(2.2.4) \qquad \chi(x,\theta,\sigma) = \varphi(x,\theta) - \psi(x,\sigma).$$

So the proof follows from an application of Proposition 2.1.1, or rather its proof, to (2.2.3). Write (2.2.3) in the form (2.2.2) to be sure of having absolutely convergent integrals and to justify the partial integrations. $\qquad\square$

2.3. Oscillatory integrals with nondegenerate phase functions

Theorem 2.3.1. *Let* a, φ *be as in Theorem 2.2.1,* $\rho > 1/2$. *Let* $\psi \in C^\infty(X \times \Sigma)$, $\xi_0 = d_x\psi(x_0,\sigma_0) \neq 0$,

$$(2.3.1) \qquad d_{(x,\theta)}[\varphi - \psi](x_0,\theta_0,\sigma_0) = 0$$

and finally

$$(2.3.2) \qquad d^2_{(x,\theta)}[\varphi - \psi](x_0,\theta_0,\sigma_0)$$

is nondegenerate.

Then there exists a neighborhood X_0 *of* x_0, Σ_0 *of* σ_0, *and a conic neighborhood* Γ_0 *of* (x_0,θ_0) *in* $X \times \mathbb{R}^N \setminus \{0\}$ *such that if* $u \in C_0^\infty(X)$, supp $u \subset X_0$, *and* ess supp $a \subset \Gamma_0$, *we have the following asymptotic development:*

$$(2.3.3) \qquad
\begin{aligned}
\langle e^{-i\tau\psi(\cdot,\sigma)}u, A \rangle &\sim e^{-i\tau\psi(x(\sigma),\sigma)} \cdot \tau^{\frac{1}{2}(N-n)} \\
&\quad \cdot (2\pi)^{\frac{1}{2}(N+n)} \cdot |\det Q(\sigma)|^{-\frac{1}{2}} \cdot e^{\frac{\pi i}{4}\operatorname{sgn} Q(\sigma)} \\
&\quad \cdot \sum_{k=0}^{\infty} \frac{1}{k!}(R(\sigma)^k g(y,\sigma,\tau))_{y=0} \cdot \tau^{-k}
\end{aligned}$$

for $\tau \to \infty$, *uniformly in* $\sigma \in \Sigma_0$.

Here $(x(\sigma),\theta(\sigma))$ *is the unique solution in* Γ_0 *of*

$$(2.3.4) \qquad d_{(x,\theta)}[\varphi - \psi](x(\sigma),\theta(\sigma),\sigma) = 0,$$

$$(2.3.5) \qquad Q(\sigma) = d^2_{(x,\theta)}[\varphi - \psi](x(\sigma),\theta(\sigma),\sigma),$$

$$(2.3.6) \qquad g(y(x,\theta,\sigma),\sigma,\tau) \cdot |\det d_{(x,\theta)}y(x,\theta,\sigma)| = a(x,\tau\theta) \cdot u(x),$$

where $(x,\theta) \to y(x,\theta,\sigma)$ *is a diffeomorphism such that*

$$(2.3.7) \qquad y(x(\sigma),\theta(\sigma),\sigma) = 0, \quad d_{(x,\theta)}y(x(\sigma),\theta(\sigma),\sigma) = I,$$

and

$$(2.3.8) \qquad [\varphi - \psi](x, \theta, \sigma) = -\psi(x(\sigma), \sigma) + \tfrac{1}{2}\langle Q(\sigma)y, y\rangle.$$

Finally we have used the second-order partial differential operator

$$(2.3.9) \qquad R(\sigma) = \tfrac{1}{2}i\Big\langle Q(\sigma)^{-1}\frac{\partial}{\partial y}, \frac{\partial}{\partial y}\Big\rangle.$$

Proof. Let $\alpha \in C_0^\infty(X \times \mathbb{R}^N \setminus \{0\})$ be equal to 1 on a neighborhood of (x_0, θ_0) and vanish outside such a small neighborhood of (x_0, θ_0) such that $\varphi - \psi$ has no critical points as a function of $(x, \theta) \in \text{supp}(1 - \alpha)$ if $\sigma \in \Sigma_0$. Then

$$\langle e^{-i\tau\psi}u, A\rangle = I_1(\tau) + I_2(\tau),$$

where

$$I_1(\tau) = \tau^N \iint e^{i\tau(\varphi-\psi)}\,\alpha a u\, dx\, d\theta$$

$$I_2(\tau) = \tau^N \iint e^{i\tau(\varphi-\psi)}(1 - \alpha)\, au\, dx\, d\theta.$$

Using the proof of Theorem 2.2.2 we see that $I_2(\tau)$ is rapidly decreasing as $\tau \to \infty$, uniformly in $\sigma \in \Sigma_0$. The integral $I_1(\tau)$ has its amplitude supported in a fixed compact subset of (x, θ)-space so (2.3.3) now immediately follows from Lemma 1.2.2 and Proposition 1.2.4. Note that $d_{(x,\theta)}(\varphi-\psi) = 0$ implies $d_\theta\varphi(x, \theta) = 0$ so $\varphi(x, \theta) = \langle\theta, d_\theta\varphi(x, \theta)\rangle = 0$ in view of the homogeneity of φ. $\qquad\square$

The asymptotic development (2.3.3) characterizes the distribution A modulo $C^\infty(X)$ in the following sense: If $\tilde{A} \in \mathcal{D}'(X)$ also satisfies (2.3.3) then $x_0 \notin \text{sing supp}\,(A - \tilde{A})$. We now analyze the conditions (2.3.1) and (2.3.2). Let Γ be a cone in $X \times \mathbb{R}^N \setminus \{0\}$. The phase function φ is called *nondegenerate in* Γ if

$$(2.3.10) \qquad d_\theta\varphi(x, \theta) = 0, \ (x, \theta) \in \Gamma \ \Rightarrow \ d_{(x,\theta)}\frac{\partial\varphi(x, \theta)}{\partial\theta_j}$$

are linearly independent for $j = 1, \ldots, N$.

The condition (2.3.10) implies that for some open cone $\tilde{\Gamma} \supset \Gamma$ the manifold

$$(2.3.11) \qquad C_\varphi = \{(x, \theta) \in \tilde{\Gamma}; \ d_\theta\varphi(x, \theta) = 0\}$$

is a conic C^∞ submanifold of $X \times \mathbb{R}^N \setminus 0$ of dimension $(n + N) - N = n$. This is a direct application of the implicit function theorem.

Lemma 2.3.2. *If φ is a nondegenerate phase function, then*

$$(2.3.12) \qquad\qquad T^{(\varphi)} : (x, \theta) \mapsto (x, d_x\varphi(x, \theta))$$

is an immersion: $C_\varphi \to T^*(X) \setminus 0$, *commuting with the multiplication with positive real numbers in the fibers. So its image Λ_φ is an immersed n-dimensional conic submanifold of $T^*(X) \setminus 0$.*

Proof. We have to prove that

$$(\delta x, \delta\theta) \in T_{(x,\theta)}(C_\varphi), \ DT^{(\varphi)}_{(x,\theta)}(\delta x, \delta\theta) = 0 \quad \Rightarrow \quad (\delta x, \delta\theta) = 0.$$

Now

$$(\delta x, \delta\theta) \in T_{(x,\theta)}(C_\varphi) \quad \Leftrightarrow \quad d_x d_\theta\varphi\, \delta x + d_\theta d_\theta\varphi\, \delta\theta = 0$$

and

$$DT^{(\varphi)}_{(x,\theta)}(\delta x, \delta\theta) = (\delta x, d_x d_x\varphi\, \delta x + d_\theta d_x\varphi\, \delta\theta)$$

so we have to prove that $d_\theta d_x\varphi\, \delta\theta = 0$, $d_\theta d_\theta\varphi\, \delta\theta = 0$ implies that $\delta\theta = 0$. But this is exactly the condition that φ is nondegenerate. $\qquad\qquad\square$

Lemma 2.3.3. *The following assertions are equivalent.*

(i) $\varphi - \psi$ *has a nondegenerate stationary point as a function of (x, θ) in* $x = x_0$, $\theta = \theta_0$, $\sigma = \sigma_0$.

(ii) a) φ *is a nondegenerate phase function in a conic neighborhood of* (x_0, θ_0) *and*

    b) *The graph of $d_x\psi$ intersects Λ_φ transversally in the point* $(x_0, \xi_0) \in T^*(X) \setminus 0$, *where*

$$\xi_0 = d_x\varphi(x_0, \theta_0) = d_x\psi(x_0, \sigma_0).$$

Proof. First note that $d_{(x,\theta)}(\varphi - \psi)(x_0, \theta_0, \sigma_0) = 0 \Leftrightarrow (x_0, d_x\psi(x_0, \sigma_0)) = (x_0, d_x\varphi(x_0, \theta_0)) \in \Lambda_\varphi$.

Second, the stationary point is nondegenerate if and only if:

$$d_x(d_x\varphi - d_x\psi) \cdot \delta x + d_\theta d_x\varphi \cdot \delta\theta = 0$$

$$(2.3.13) \qquad\qquad \text{and}$$

$$d_x d_\theta\varphi \cdot \delta x + d_\theta d_\theta\varphi \cdot \delta\theta = 0$$

imply that $\delta x = 0$, $\delta\theta = 0$. In particular (2.3.13), taking $\delta x = 0$, implies

$$d_\theta d_x\varphi \cdot \delta\theta = 0, \ d_\theta d_\theta\varphi \cdot \delta\theta = 0 \quad \Rightarrow \quad \delta\theta = 0,$$

that is, φ is a nondegenerate phase function.

If conversely φ is a nondegenerate phase function, then Λ_φ has tangent space in $(x_0, d_x\varphi(x_0, \theta_0))$ equal to

$$\{(\delta x, d_x d_x\varphi \cdot \delta x + d_\theta d_x\varphi \cdot \delta\theta); \ d_x d_\theta\varphi \cdot \delta x + d_\theta d_\theta\varphi \cdot \delta\theta = 0\}.$$

The tangent space to the graph of $d_x\psi$ is equal to $\{(\delta x, d_x d_x\psi \cdot \delta x); \ \delta x \in T_{x_0}(X)\}$. The condition (2.3.13) means precisely that these spaces have zero intersection, that is, are transversal. (Note that $d\psi$ and Λ_φ have dimension equal to n and $\dim T^*(X) = 2n$.) $\qquad\square$

Lemma 2.3.3 implies that if φ is a nondegenerate phase function on ess supp a, then the condition (2.3.2) is satisfied for the general function ψ satisfying (2.3.1). Indeed for any $(x_0, \xi_0) \in T^*(X)$ and any symmetric $n \times n$ matrix B there is a C^∞ function ψ on X such that $d\psi(x_0) = \xi_0$, $d^2\psi(x_0) = B$. Here we use that the set of symmetric B such that $\{(\delta x, B\delta x); \ \delta x \in T_{x_0}(X)\}$ is transversal to $T_{(x_0, \xi_0)}(\Lambda_\varphi)$ is open and dense in the space of all symmetric B. (See Theorem 3.3.7.)

From (2.3.3) we see therefore that if $\varphi(x, \theta)$ and $\tilde\varphi(x, \tilde\theta)$ are nondegenerate phase functions at $(x_0, \theta_0) \in X \times \mathbb{R}^N \setminus \{0\}$ and $(x_0, \tilde\theta_0) \in X \times \mathbb{R}^{\tilde N} \setminus \{0\}$, respectively, then any distribution A, defined by the phase function φ and some amplitude a such that ess supp a is contained in a conic neighborhood of (x_0, θ_0), is modulo $C^\infty(X)$ equal to a distribution defined by the phase function $\tilde\varphi$ and some other amplitude $\tilde a$, only if the corresponding manifolds Λ_φ, $\Lambda_{\tilde\varphi}$ coincide. The following theorem shows that also the converse is true.

Theorem 2.3.4. *Suppose $\varphi(x, \theta)$ and $\tilde\varphi(x, \tilde\theta)$ are nondegenerate phase functions at $(x, \theta_0) \in X \times \mathbb{R}^N \setminus \{0\}$ and at $(x_0, \tilde\theta_0) \in X \times \mathbb{R}^{\tilde N} \setminus \{0\}$, respectively. Let Γ and $\tilde\Gamma$ be open conic neighborhoods of (x_0, θ_0) and $(x_0, \tilde\theta_0)$ such that $T_\varphi \colon C_\varphi \to \Lambda_\varphi$ and $T_{\tilde\varphi} \colon C_{\tilde\varphi} \to \Gamma_{\tilde\varphi}$ are injective, respectively. If $\Lambda_\varphi = \Lambda_{\tilde\varphi}$ then any Fourier integral A, defined by the phase function φ and an amplitude $a \in S_\rho^\mu(X \times \mathbb{R}^N)$, $\rho > 1/2$, with ess supp a contained in a sufficiently small conic neighborhood of (x_0, θ_0), is equal to a Fourier integral defined by the phase function $\tilde\varphi$ and an amplitude $\tilde a \in S_\rho^{\mu + \frac{1}{2}(N - \tilde N)}(X \times \mathbb{R}^{\tilde N})$.*

Proof. We start by reducing the number of θ-variables as far as possible. Let $(x_0, \theta_0) \in \Gamma$. Applying an orthogonal transformation in the θ-variables we can assume that

$$\theta = (\theta', \theta''), \quad \theta' = (\theta_1, \ldots, \theta_k), \quad \theta'' = (\theta_{k+1}, \ldots, \theta_N),$$

(2.3.14) $$\qquad d^2_{\theta'} \varphi(x_0, \theta_0) = 0, \quad d_{\theta'} d_{\theta''} \varphi(x_0, \theta_0) = 0,$$

and finally $d^2_{\theta''} \varphi(x_0, \theta_0)$ is nondegenerate.

Without loss of generality we may also assume that $\theta''_0 = 0$. (2.3.14) implies that we can solve $\theta'' = \theta''(x, \theta')$ from the equation $d_{\theta''} \varphi(x, \theta', \theta'') = 0$, $\theta''(x_0, \theta'_0) = 0$. Write $\varphi(x, \theta', \theta'') = \varphi_1(x, \theta') + \psi(x, \theta', (\psi(x, \theta', \theta'')) \theta''$ where $\varphi_1(x; \theta') = \varphi(x, \theta', \theta''(x, \theta'))$. With these notations, we now write

(2.3.15)
$$\langle u, A \rangle = \iiint e^{i\varphi(x, \theta', \theta'')} a(x, \theta', \theta'') \, u(x) \, dx \, d\theta' \, d\theta''$$
$$= \iint e^{i\varphi_1(x, \theta')} b(x, \theta') \, u(x) \, dx \, d\theta',$$

where

(2.3.16) $$\qquad b(x, \theta') = \int e^{i\psi(x, \theta', \theta'')} a(x, \theta', \theta'') \, d\theta''.$$

Here $d^2_{\theta''} \psi(x_0, \theta_0)$ is nondegenerate, so if $a \in S^\mu_\rho(X \times \mathbb{R}^N)$, $\rho > 1/2$, and ess supp a contained in a sufficiently small conic neighborhood of (x_0, θ_0) we see from Section 1.2 that $b \in S^{\mu + \frac{1}{2}(N-k)}_\rho(X \times \mathbb{R}^k)$.

Now $d_{\theta'} d_{\theta''} \varphi(x_0, \theta_0) = 0$ implies that φ_1 is a nondegenerate phase function on a conic neighborhood of (x_0, θ'_0) and $\Lambda_{\phi_1} = \Lambda_\phi$ (locally).

$d^2_{\theta'} \varphi(x_0, \theta_0) = 0$ implies that $\{x_0\} \times \mathbb{R}^{N-k}$ is contained in $T_{(x_0, \theta'_0)}(C_{\varphi_1})$, so the differential of the projection $C_{\varphi_1} \ni (x, \theta') \to x$ has in (x_0, θ'_0) rank equal to $n - k$. This rank is equal to the rank of the projection $\Lambda_{\varphi_1} \ni (x, \xi) \to x \in X$ in (x_0, ξ_0). So the kernel of the latter projection has rank equal to k and we have proved

Lemma 2.3.5. *The number of θ-variables is $\geq$ the dimension k of the intersection of the tangent spaces of Λ_φ, and the fiber of $T^*(X)$, at (x_0, ξ_0). Moreover, every Fourier integral defined by the phase function φ and amplitude $a \in S^\mu_\rho(X \times \mathbb{R}^N)$, ess supp a in a sufficiently small conic neighborhood of (x_0, θ_0), can also be defined by a phase function in k variables and amplitude $b \in S^{\mu + \frac{1}{2}(N-k)}_\rho(X \times \mathbb{R}^k)$ given by (2.3.16).*

Conversely we can always raise the number of frequency variables from k to an arbitrary $\tilde{N} \geq k$, because any $b \in S^{\mu + \frac{1}{2}(N-k)}_\rho(X \times \mathbb{R}^k)$ arises by (2.3.16) from some $\tilde{a} \in S^{\mu + \frac{1}{2}(N-\tilde{N})}(X \times \mathbb{R}^{\tilde{N}})$, if we replace φ by $\tilde{\varphi}$ in the definition preceding (2.3.16). So we now may assume that $N = \tilde{N} \; (= k)$. In view of the formula

$$\iint e^{i\tilde{\varphi}(x,\tilde{\theta})} a(x,\tilde{\theta}) u(x) \, dx \, d\tilde{\theta} =$$

(2.3.17)

$$\iint e^{i\varphi(x,\theta)} a(x,\tilde{\theta}(x,\theta)) \cdot |\det d_\theta \tilde{\theta}(x,\theta)| \cdot u(x) \, dx \, d\theta,$$

the proof of Theorem 2.3.4 is therefore completed by

Lemma 2.3.6. *If* $\Lambda_\varphi = \Lambda_{\tilde{\varphi}} = \Lambda$, $N = \tilde{N} = minimal$, *then there exists a* C^∞ *mapping* $(x,\theta) \to \tilde{\theta}(x,\theta)$: $\Gamma \to \mathbb{R}^{\tilde{N}} \setminus \{0\}$, *homogeneous in* θ *of degree* 1, *such that*

(2.3.18) $\qquad \tilde{\theta}(x_0, \theta_0) = \tilde{\theta}_0, \qquad \tilde{\varphi}(x, \tilde{\theta}(x,\theta)) = \varphi(x,\theta)$ *on* Γ.

Proof of Lemma 2.3.6. The diffeomorphisms $T_\varphi : C_\varphi \to \Lambda$, $T_{\tilde{\varphi}} : C_{\tilde{\varphi}} \to \Lambda$ induce a diffeomorphism $T := T_{\tilde{\varphi}}^{-1} \circ T_\varphi : \; C_\varphi \to C_{\tilde{\varphi}}$. If π denotes the projection $(x,\theta) \mapsto x$, then we have $\pi(T(x,\theta)) = x$ on C_φ, so $T(x,\theta) = (x, \tilde{\theta}(x,\theta))$ for some C^∞ function $\tilde{\theta}(x,\theta)$ on C_φ. $\tilde{\theta}$ is homogeneous in θ of degree 1 because T_φ and $T_{\tilde{\varphi}}$ are, and we have

$$\tilde{\varphi}(x, \tilde{\theta}(x,\theta)) = \varphi(x,\theta) = 0,$$

(2.3.19)

$$d_{\tilde{\theta}} \tilde{\varphi}(x, \tilde{\theta}(x,\theta)) = d_\theta(x,\theta) = 0,$$

$$d_x \tilde{\varphi}(x, \tilde{\theta}(x,\theta)) = d_x \varphi(x,\theta), \quad \text{on } C_\varphi.$$

Now extend $\tilde{\theta}$ to a homogeneous C^∞ function of degree 1 on a conic neighborhood in $X \times (\mathbb{R}^N \setminus \{0\})$ of C_φ, define $\psi(x,\theta) = \tilde{\varphi}(x, \tilde{\theta}(x,\theta))$ there. Then ψ is a nondegenerate phase function, $\Lambda_\psi = \Lambda$ and $\psi - \varphi$ vanishes of order 2 on $C_\varphi = C_\psi$. We are ready if we can find a homogeneous C^∞ function $\theta'(x,\theta)$ of degree 1 such that $\psi(x, \theta'(x,\theta)) = \varphi(x,\theta)$.

Because C_ψ is equal to the manifold $d_\theta \psi = 0$ and we can choose $\frac{\partial \psi}{\partial \theta_1}, \ldots, \frac{\partial \psi}{\partial \theta_N}$ as the first N coordinates in a coordinatization, Taylor development at C_ψ gives

(2.3.20) $\qquad (\psi - \varphi)(x,\theta) = B(x,\theta)(d_\theta \psi(x,\theta), d_\theta \psi(x,\theta))$

for some symmetric bilinear form $B(x,\theta)$, depending C^∞ on (x,θ) and being homogeneous of degree 1 in θ. Analogously

(2.3.21) $\qquad \psi(x, \theta + \mu) = \psi(x,\theta) + \langle d_\theta \psi(x,\theta), \mu \rangle + C(x,\theta)(\mu, \mu)$

for some other symmetric bilinear form $C(x, \theta)$, homogeneous in θ of degree -1.

Now try $\theta'(x, \theta) = \theta + W(x, \theta) d_\theta \psi(x, \theta)$ for some W. This leads to the equation

$$(2.3.22) \qquad\qquad W + W'CW = B$$

on C_ψ. This equation has the solution $W = 0$ if $B = 0$, so the implicit function theorem gives a unique C^∞ solution $W(x, \theta)$ on a conic neighborhood of C_ψ, homogeneous in θ of degree 1. Note that $B(x_0, \theta_0) = 0$ because at (x_0, θ_0) we have $d_\theta^2 \varphi = d_\theta^2 \psi = 0$, which also implies that $d_x d_\theta \varphi = d_x d_\theta \psi$.   □

The above theorems show that one should rather speak of distributions A *defined by a conic manifold* Λ in $T^*(X) \setminus 0$, which locally is equal to Λ_φ, φ a nondegenerate phase function, instead of distributions defined by some phase function φ. A differential geometric characterization of such manifolds Λ will be given in Chapter 3, after we have studied the differential geometric structure of the cotangent bundle $T^*(X)$ in more detail. After Chapter 3 we will also be able to give an invariant (that is, independent of the choice of the phase function) characterization of the principal symbol of A. In a primitive way the principal symbol of A at (x_0, ξ_0) can be defined as the mapping assigning to each $\psi \in C^\infty(X)$, with $\xi_0 = d_x \psi(x_0) \neq 0$ the top order term in the asymptotic development of $e^{i\tau\psi(x_0)} \cdot \langle e^{-i\tau\psi} u, A \rangle$ for $\tau \to \infty$, that is, the quantity

$$(2.3.23) \qquad \begin{aligned} &\tau^{\frac{1}{2}(N-n)}(2\pi)^{\frac{1}{2}(N+n)} \cdot |\det Q|^{-\frac{1}{2}} \cdot e^{\frac{\pi i}{4} \operatorname{sgn} Q} \\ &\qquad \cdot a(x_0, \tau\theta_0) \cdot u(x_0) + O(\tau^{\mu + \frac{1}{2}(N-n) + (1-2\rho)}) \text{ for } \tau \to \infty. \end{aligned}$$

Since the dependence on ψ only involves $d_x^2 \psi(x_0)$ we see that the principal symbol s is a function on the space of symmetric matrices B such that $\{(\delta x, B \cdot \delta x)\}$ is transversal to $T_{(x_0, \xi_0)}(\Lambda_\varphi)$. We see that

$$(2.3.24) \qquad\qquad s(B') = t(B', B) \cdot s(B)$$

where the factor $t(B', B)$ does not depend on the amplitude a. The collection of all functions s satisfying (2.3.24) therefore is a complex one-dimensional space $L_\varphi(x_0, \xi_0)$, which only depends on Λ_φ. The $L_\varphi(x_0, \xi_0)$, $(x_0, \xi_0) \in \Lambda_\varphi$ form a complex line bundle L over Λ_φ of which the principal symbol of A becomes a section. Also this complex line bundle L can only be understood better after we have obtained more insight into the differential geometric structure of $T^*(X)$.

2.4. Fourier integral operators (local theory)

A *Fourier integral operator* is defined as an operator $A \colon C_0^\infty(Y) \to \mathcal{D}'(X)$ such that the distribution kernel $K_A \in \mathcal{D}'(X \times Y)$ is a Fourier integral defined by a nondegenerate phase function in an open cone Γ in $T^*(X \times Y) \setminus 0$ and some amplitude $a \in S_\rho^\mu(X \times Y \times \mathbb{R}^N)$, ess supp $a \subset \Gamma$. We have $WF'(A) \subset \Lambda'_\varphi$ if we write

$$(2.4.1) \qquad \Lambda' = \big\{ ((x,\xi),(y,\eta)) \in T^*(X) \times T^*(Y); (x,y,\xi,-\eta) \in \Lambda \big\}.$$

for any subset Λ of $T^*(X \times Y)$. Conversely each $(x,\xi) \in \Lambda'_\varphi$ can be made to be an element of $WF'(A)$ by a suitable choice of the amplitude, using Theorem 2.3.1.

According to Theorem 1.4.1 the operator A can be extended to $\mathcal{D}'_V(Y) \cap \mathcal{E}'(Y)$ (and to $\mathcal{D}'_V(Y)$ if suitable assumptions are made on supp K_A) if there are no $((x,\xi),(y,\eta)) \in \Lambda'_\varphi$ such that $\xi = 0$ and $(y,\eta) \in V$. In particular A can be extended to $\mathcal{E}'(Y)$ (and to $\mathcal{D}'(Y)$, respectively) if $d_x\varphi \neq 0$ when $d_\theta\varphi = 0$, that is, if φ has no stationary points as a function of (x,θ). On the other hand A maps $C_0^\infty(Y)$ into $C^\infty(X)$ if φ has no stationary points as a function of (y,θ). Note that it was included in the definition of a phase function that φ has no stationary points as a function of (x,y,θ).

Examples (more will follow in Chapters 4 and 5).

(1) Fourier integrals in X can be regarded as Fourier integral operators by taking $Y = \{\text{point}\}$.

(2) Let κ be a smooth map from X to Y. Then

$$(\kappa^* u)(x) = u(\kappa(x)) = (2\pi)^{-n} \iint e^{i\langle \kappa(x)-y,\eta \rangle} u(y)\, dy\, d\eta,$$

and it follows that $\kappa^* \colon C^\infty(Y) \to C^\infty(X)$ is a Fourier integral operator defined by a nondegenerate phase function φ such that

$$(2.4.2) \qquad \Lambda'_\varphi = \big\{ ((x,\xi),(y,\eta)); y = \kappa(x), \xi = {}^tD\kappa_x \cdot \eta \big\}.$$

If κ is a diffeomorphism, then Λ'_φ is the graph of the induced transformation $\tilde\kappa \colon T^*(X) \setminus 0 \to T^*(Y) \setminus 0$ defined by

$$(2.4.3) \qquad \tilde\kappa(x,\xi) = (\kappa(x), ({}^tD\kappa_x)^{-1}(\xi)).$$

If X is a submanifold of Y, $\dim X < \dim Y$ and κ is the identity: $X \to Y$ then κ^* is the restriction operator $\rho : C^\infty(Y) \to C^\infty(X)$. In this case

$$(2.4.4) \qquad \Lambda'_\varphi = \{((x,\xi),(y,\eta)); y = x, \xi = \eta|_{T_x(X)}\}$$

which is far from the graph of a map. ρ can be extended continuously to $\mathcal{D}'_\Gamma(Y)$, for any closed cone Γ in $T^*(Y) \setminus 0$ that does not meet the set

$$(2.4.5) \qquad \{(y,\eta) \in T^*(Y) \setminus 0; y \in X, \eta|_{T_x(X)} = 0\},$$

that is, the *normal bundle in* $T^*(Y) \setminus 0$ *of the submanifold* X.

(3) *Pseudodifferential operators in* X are defined as Fourier integral operators with $Y = X$ and

$$
\begin{aligned}
\Lambda'_\varphi &\subset \text{diagonal in } T^*(X) \setminus 0 \times T^*(X) \setminus 0 \\
(2.4.6) \qquad &= \text{graph of the identity: } T^*(X) \setminus 0 \to T^*(X) \setminus 0.
\end{aligned}
$$

These operators will be considered in more detail in Section 2.5.

In view of Theorem 1.3.7 the following product theorem seems the natural one. We denote

$$(2.4.7) \qquad \operatorname{diag} V = \{(v,v) \in V \times V; v \in V\}$$

for any set V.

· **Theorem 2.4.1.** *Let* X, Y, Z *be open in* $\mathbb{R}^{n_X}$, $\mathbb{R}^{n_Y}$, $\mathbb{R}^{n_Z}$ *respectively. Let* A_1 *be a Fourier integral operator:* $C_0^\infty(Y) \to \mathcal{D}'(X)$ *defined by a nondegenerate phase function* φ_1 *in an open cone* Γ_1 *in* $X \times Y \times \mathbb{R}^{N_1} \setminus \{0\}$ *and an amplitude* $a_1 \in S^{\mu_1}_\rho(X \times Y \times \mathbb{R}^{N_1})$, *ess supp* $a_1 \subset \Gamma_1$. *Similarly* A_2 *is a Fourier integral operator:* $C_0^\infty(Z) \to \mathcal{D}'(Y)$ *defined by a nondegenerate phase function* φ_2 *in an open cone* Γ_2 *in* $Y \times Z \times \mathbb{R}^{N_2} \setminus \{0\}$ *and an amplitude* $a_2 \in S^{\mu_2}_\rho(Y \times Z \times \mathbb{R}^{N_2})$, *ess supp* $a_2 \subset \Gamma_2$. *Assume that* $\rho > 1/2$ *and:*

$$(2.4.8) \qquad
\begin{aligned}
&\text{The projection from } \pi_{X \times Y}(\operatorname{supp} a_1) \times \pi_{Y \times Z}(\operatorname{supp} a_2) \\
&\cap X \times (\operatorname{diag} Y) \times Z \text{ into } X \times Z \text{ is a proper mapping,}
\end{aligned}
$$

$$(2.4.9) \qquad \eta \neq 0 \text{ if } (x,\xi,y,\eta) \in \Lambda'_{\varphi_1} \text{ or } (y,\eta,z,\xi) \in \Lambda'_{\varphi_2},$$

$$(2.4.10) \qquad \xi \neq 0 \text{ or } \zeta \neq 0 \text{ if } (x,\xi,y,\eta) \in \Lambda'_{\varphi_1} \text{ and } (y,\eta,z,\zeta) \in \Lambda'_{\varphi_2}$$

(2.4.11) $\Lambda'_{\varphi_1} \times \Lambda'_{\varphi_2}$ *intersects* $T^*(X) \times (\text{diag } T^*(Y)) \times T^*(Z)$ *transversally.*

Then $A_1 \circ A_2$ *is well defined and modulo an operator with* C^∞ *kernel equal to a Fourier integral operator:* $C_0^\infty(Z) \to \mathcal{D}'(X)$ *defined by a nondegenerate phase function* φ *in an open cone* Γ *in* $X \times Z \times \mathbb{R}^N \setminus \{0\}$, $N = N_1 + N_2 + n_Y$, *and an amplitude* $a \in S_\rho^\mu(X \times Z \times \mathbb{R}^N)$, $\mu = \mu_1 + \mu_2 - n_Y$, *ess supp* $a \subset \Gamma$. *Moreover,*

(2.4.12) $$\Lambda'_\varphi = \Lambda'_{\varphi_1} \circ \Lambda'_{\varphi_2}.$$

Proof. $A_1 \circ A_2$ is well defined in view of (2.4.8), (2.4.9) and Theorem 1.4.1, and we have

(2.4.13)
$$K_{A_1 \circ A_2}(x,z) = \iiint e^{i[\varphi_1(x,y,\theta) + \varphi_2(y,z,\sigma)]} a_1(x,y,\theta)$$
$$\cdot a_2(y,z,\sigma)\, dy\, d\theta\, d\sigma$$

in the distribution sense. If we let x, z vary in a compact subset of $X \times Z$ then the corresponding $(x, z, \theta, \sigma, y)$ such that $|(\theta, \sigma)| = (|\theta|^2 + |\sigma|^2)^{1/2} = 1$, $(x, y, \theta) \in \text{supp } a_1$, $(y, z, \sigma) \in \text{supp } a_2$, vary in a compact set in view of (2.4.8). What follows will only refer to such $(x, z, \theta, \sigma, y)$.

Because $d_{(y,\theta)} \varphi_1(x, y, \theta) \neq 0$ if $\theta \neq 0$ in view of (2.4.9) there exists $0 < \varepsilon < 1$ such that $|\sigma| \leq \varepsilon |\theta|$,

$$|(\theta, \sigma)| = 1 \Rightarrow d_{(y,\theta)}[\varphi_1 + \varphi_2](x, z, \theta, \sigma, y) \neq 0.$$

Let χ_1 be a homogeneous C^∞ function of degree 0 in (θ, σ) such that $\chi_1 = 1$ for $|\sigma| \leq \frac{1}{2}\varepsilon|\theta|$ and $\chi_1 = 0$ for $|\sigma| > \varepsilon|\theta|$ if $|(\theta, \sigma)| = 1$. It follows, using partial integration in (x, θ) and using the bound $|\sigma| \leq \varepsilon|\theta|$ for σ, that

(2.4.14) $$\iiint e^{i(\varphi_1 + \varphi_2)} \chi_1 \cdot a_1 \cdot a_2 \, dy\, d\theta\, d\sigma$$

is a C^∞ function of (x, z).

Similarly there is a homogeneous C^∞ function of degree 0 in (θ, σ) such that $\chi_2 = 1$ for $|\theta| \leq \frac{1}{2}\varepsilon|\sigma|$, $\chi_2 = 0$ for $|\theta| > \varepsilon|\sigma|$ if $|(\theta, \sigma)| = 1$ and

(2.4.15) $$\iiint e^{i(\varphi_1 + \varphi_2)} \chi_2 \cdot a_1 \cdot a_2 \, dy\, d\theta\, d\sigma$$

is a C^∞ function of (x, z). So we are left with the integral

(2.4.16) $$\iiint e^{i[\varphi_1 + \varphi_2]} b \, dy\, d\theta\, d\sigma, \quad \text{where } b = (1 - \chi_1 - \chi_2)a_1 a_2.$$

Notice that

(2.4.17) $\frac{1}{2}\varepsilon|\sigma| \le |\theta| \le 2\varepsilon^{-1}|\sigma|$ on supp b.

We claim that (2.4.16) is a Fourier integral with phase

(2.4.18) $\varphi(x, z, (\theta, \sigma, \tilde{y})) = \varphi_1(x, \tilde{y}/|(\theta, \sigma)|, \theta) + \varphi_2(\tilde{y}/|(\theta, \sigma)|, z, \sigma)$

(frequency variables $(\theta, \sigma, \tilde{y})$), and amplitude

(2.4.19) $a(x, z, (\theta, \sigma, \tilde{y})) = b(x, z, \theta, \sigma, \tilde{y}/|(\theta, \sigma)|) \cdot |(\theta, \sigma)|^{-n_Y}.$

Indeed, because of (2.4.17) we stay away from the boundaries $\theta = 0$, $\sigma \ne 0$ and $\theta \ne 0$, $\sigma = 0$ of the region where φ is a C^∞ function. Moreover (2.4.10) implies that $d_x\varphi_1 \ne 0$ or $d_z\varphi_2 \ne 0$ if $d_\theta\varphi_1 = 0$, $d_\sigma\varphi_2 = 0$, $d_y(\varphi_1 + \varphi_2) = 0$, so φ is a phase function as defined in Section 2.2.

Inequality (2.4.17) also implies that $a \in S_\rho^\mu(X \times Z \times \mathbb{R}^N)$. Indeed, derivation with respect to, say, θ improves by a factor $(1 + |\theta|)^{-\rho}$, which in fact is an improvement by a factor $(1 + |\theta| + |\sigma|)^{-\rho}$ in view of $|\theta| \ge \frac{1}{2}\varepsilon|\sigma|$.

We now investigate (2.4.11) and (2.4.12). $\Lambda'_{\varphi_1} \times \Lambda'_{\varphi_2}$ intersects $T^*(X) \times$ (diag $T^*(Y)) \times T^*(Z)$ precisely at the points $(x, d_x\varphi_1, y, -d_y\varphi_1, y, d_y\varphi_2, z, -d_z\varphi_2)$ where $-d_y\varphi_1 = d_y\varphi_2$, $d_\theta\varphi_1 = 0$, $d_\theta\varphi_2 = 0$, which proves (2.4.12).

The tangent space of Λ'_{φ_1} consists of the vectors

(2.4.20)
$$(\delta x, d(d_x\varphi_1)u, \ \delta y, \ -d(d_y\varphi_1)u),$$
$$\text{such that} \quad d(d_\theta\varphi_1)u = 0, \quad u = (\delta x, \delta y, \delta\theta),$$

and the tangent space of Λ'_{φ_2} consists of the vectors

(2.4.21)
$$(\delta y, d(d_y\varphi_2)v, \ \delta z, \ -d(d_z\varphi_2)v),$$
$$\text{such that} \quad d(d_\sigma\varphi_2)v = 0, \quad v = (\delta y, \delta z, \delta\sigma),$$

The intersection of $T(\Lambda'_{\varphi_1} \times \Lambda'_{\varphi_2}) = T(\Lambda'_{\varphi_1}) \times T(\Lambda'_{\varphi_2})$ with $T(T^*(X) \times$ (diag $T^*(Y)) \times T^*(Z))$ therefore has the same dimension as the kernel of $d\, d_{(\theta,\sigma,y)}[\varphi_1 + \varphi_2]$. The intersection is transversal if and only if this dimension is equal to

$$((n_X + n_Y) + (n_Y + n_Z)) + (2n_X + 2n_Y + 2n_Z)$$
$$- (2n_X + 4n_Y + 2n_Z) = n_X + n_Z,$$

that is, if and only if rank $d\, d_{(\theta,\sigma,y)}[\varphi_1 + \varphi_2] = N_1 + N_2 + n_Y$. But this means precisely that φ is a nondegenerate phase function. $\square$

We conclude this section by a little philosophy on the question: what should we call the *order* of a Fourier integral operator defined by a phase φ and an amplitude $a \in S^\mu_\rho(X \times Y \times \mathbb{R}^N)$? Suppose that the order is a function of μ, $n_X + n_Y$, and N, and suppose that the order of a product (when defined) is equal to the sum of the orders. In view of Lemma 2.3.5 it must be of the form $\mu + \frac{1}{2}N + f(n_X + n_Y)$, and the additivity of the orders means that

$$f(n_X + n_Z) = f(n_X + n_Y) + f(n_Y + n_Z) + \tfrac{1}{2}n_Y$$

in view of Theorem 2.4.1. So we are led to define the order of a Fourier integral operator $A: C_0^\infty(Y) \to \mathcal{D}'(X)$ defined by an amplitude of order μ as

(2.4.22) $$\operatorname{order}(A) = \mu + \tfrac{1}{2}N - \tfrac{1}{4}(n_X + n_Y).$$

With this choice Theorem 2.4.1 can be supplemented by:

(2.4.23) $$\operatorname{order}(A_1 \circ A_2) = \operatorname{order} A_1 + \operatorname{order} A_2.$$

2.5. Pseudodifferential operators in $\mathbb{R}^n$

A *pseudodifferential operator of order m and type ρ* on $\mathbb{R}^n$ is defined as a Fourier integral operator A with phase

(2.5.1) $$\varphi(x, y, \eta) = \langle x - y, \eta \rangle,$$

and amplitude $a(x, y, \eta) \in S^m_\rho((\mathbb{R}^n \times \mathbb{R}^n) \times \mathbb{R}^n)$. Observe that the order m coincides with the order defined at the end of Section 2.4. Writing $X = \mathbb{R}^n$, we see that $\Lambda'_\varphi = \operatorname{diag} T^*(X) \setminus 0 = $ graph of the identity: $T^*(X) \setminus 0 \to T^*(X) \setminus 0$, and in view of Theorem 2.3.4 we could replace (2.5.1) by any nondegenerate phase function $\varphi(x, y, \theta)$ on $\mathbb{R}^n \times \mathbb{R}^n \times \mathbb{R}^N$ such that $d_\theta \varphi(x, y, \theta) = 0 \Rightarrow x = y$, $d_x \varphi(x, y, \theta) = -d_y \varphi(x, y, \theta)$.

Because of $\Lambda'_\varphi = \operatorname{diag} T^*(X) \setminus 0$ we can write

(2.5.2) $$WF'(A) = \operatorname{diag} WF(A)$$

for a uniquely defined closed conic subset $WF(A)$ of $T^*(\mathbb{R}^n) \setminus 0$, called the *wave front set of the pseudodifferential operator A* (by abuse of language). The space of all pseudodifferential operators of order m and type ρ will be denoted by $L^m_\rho(\mathbb{R}^n)$.

Definition 2.5.1. Let $A \in L_\rho^m(\mathbb{R}^n)$, $\rho > 1/2$, be properly supported. The *complete symbol* σ_A of A is defined by

$$(2.5.3) \qquad \sigma_A(x, \eta) = e^{-i\langle x, \eta \rangle} A(e^{i\langle \cdot, \eta \rangle})(x).$$

We now compare this with the definition of the symbol in the introduction:

Theorem 2.5.1. *Let A be as in Definition 2.5.1. Then*

$$(2.5.4) \qquad (Au)(x) = (2\pi)^{-n} \iint e^{i\langle x-y, \eta \rangle} \sigma_A(x, \eta) u(y) \, dy \, d\eta$$

for all $u \in C_0^\infty(\mathbb{R}^n)$. If A is given by the amplitude $a(x, y, \eta)$, then

$$(2.5.5) \qquad \sigma_A(x, \eta) \sim \sum_{k=0}^{\infty} \frac{1}{k!} \left(\sum_{j=1}^{n} \frac{1}{i} \frac{\partial^2}{\partial y_j \partial \eta_j} \right)^k a(x, y, \eta) \Big|_{y=x}.$$

Proof. Write $u(y) = (2\pi)^{-n} \int e^{i\langle y, \eta \rangle} \cdot (\mathcal{F}u)(\eta) d\eta$. In view of the continuity of A this implies

$$(Au)(x) = (2\pi)^{-n} \int A(e^{i\langle \cdot, \eta \rangle})(x) \cdot (\mathcal{F}u)(\eta) \, d\eta,$$

that is, (2.5.4). For (2.5.5) we remark that

$$(Ae^{i\langle \cdot, \eta \rangle})(x) = (2\pi)^{-n} \iint e^{i\langle x-y, \theta \rangle} a(x, y, \theta) e^{i\langle y, \theta \rangle} \, dy \, d\theta$$

$$= e^{i\langle x, \eta \rangle} (2\pi)^{-n} \iint e^{-i\langle y, \theta \rangle} a(x, x+y, \eta+\theta) \, dy \, d\theta.$$

Now apply Proposition 1.2.4. $\qquad \qquad \square$

Theorem 2.5.2. *Let $A_1 \in L_\rho^{m_1}(\mathbb{R}^n)$, $A \in L_\rho^{m_2}(\mathbb{R}^n)$ be properly supported, $\rho > 1/2$. Then $A_1 \circ A_2 \in L_\rho^{m_1+m_2}(\mathbb{R}^n)$ is properly supported, and its symbol is given by*

$$(2.5.6) \qquad \sigma_{A_1 \circ A_2}(x, \zeta) \sim \sum_{k=0}^{\infty} \frac{1}{k!} \left(\sum_{j=1}^{n} \frac{1}{i} \frac{\partial^2}{\partial y_j \partial \eta_j} \right)^k \sigma_{A_1}(x, \eta) \cdot \sigma_{A_2}(y, \zeta),$$

the differentiations taken at $\eta = \zeta$, $y = x$.

Proof. Apply the reduction of frequency variables of Lemma 2.3.5 to the product formula (2.4.13). $\qquad \qquad \square$

Corollary 2.5.3. *If we define the principal symbol a of A by $a = \sigma_A$ mod $S_\rho^{m+(1-2\rho)}(\mathbb{R}^n)$, then the principal symbol of $A_1 \circ A_2$ is equal to the product of the principal symbols a_1 and a_2 of A_1 and A_2, respectively.*

Moreover, $[A_1, A_2] = A_1 \circ A_2 - A_2 \circ A_1 \in L_\rho^{m_1+m_2+1-2\rho}(\mathbb{R}^n)$ and its principal symbol of order $m_1 + m_2 + 1 - 2\rho$ is equal to $\frac{1}{i}\{a_1, a_2\}$. Here the Poisson brackets are defined by

$$(2.5.7) \qquad \{f, g\} = \sum \frac{\partial f}{\partial \xi_j} \frac{\partial g}{\partial x_j} - \sum \frac{\partial f}{\partial x_j} \frac{\partial g}{\partial \xi_j}$$

for any differentiable functions f, g on $\mathbb{R}^n \times \mathbb{R}^n = T^*(\mathbb{R}^n)$.

(Such a relation between operators on X and functions on $T^*(X)$, assigning to the commutator of the operators $\frac{1}{i} \times$ Poisson brackets of the functions, is familiar in quantum mechanics.)

Theorem 2.5.4. *If $A \in L_\rho^m(\mathbb{R}^n)$ then A is continuous: $H_{\text{comp}}^s(\mathbb{R}^n) \to H_{\text{loc}}^{s-m}(\mathbb{R}^n)$ for all $s \in \mathbb{R}$.*

Proof. $H_{\text{loc}}^s(\mathbb{R}^n)$ is the set of $u \in \mathcal{D}'(\mathbb{R}^n)$ such that $(1 + |D|^2)^{s/2}(\varphi u) \in L^2(\mathbb{R}^n)$ for all $\varphi \in C_0^\infty(\mathbb{R}^n)$. We may assume that A is compactly supported. Let $u \in H_{\text{comp}}^s(\mathbb{R}^n)$, then $(1 + |D|^2)^{s/2}u \in L_{\text{loc}}^2(\mathbb{R}^n)$, so

$$(1 + |D|^2)^{(s-m)/2}Au = (1 + |D|^2)^{(s-m)/2}A(1 + |D|^2)^{-s/2}$$
$$\cdot (1 + |D|^2)^{s/2}u \in L_{\text{loc}}^2(\mathbb{R}^n).$$

Here $B = (1 + |D|^2)^{(s-m)/2}A(1 + |D|^2)^{-s/2}$ is of order 0 and such operators map $L_{\text{comp}}^2(\mathbb{R}^n)$ into $L_{\text{loc}}^2(\mathbb{R}^n)$.

A proof of the latter statement can be given as follows. One first verifies that if $P \in L_\rho^m(\mathbb{R}^n)$, then $P^* \in L_\rho^m(\mathbb{R}^n)$, if P^* denotes the adjoint of P with respect to the L^2 inner product. This follows from Theorem 2.5.1, in combination with the observation that P^* is defined by the amplitude

$$a(x, y, \eta) = \overline{\sigma_P(y, \eta)}.$$

(One can also use Theorem 4.4.1 in combination with the definition of $L_\rho^m(X)$ following Proposition 4.2.4.)

Next, $B \in L_\rho^0(\mathbb{R}^n)$ implies that $B^*B \in L_\rho^0(\mathbb{R}^n)$, which has a bounded principal symbol. This means that there is a positive constant C, a pseudo-differential operator $P \in L_\rho^0(\mathbb{R}^n)$, and an integral operator R with smooth kernel, such that $B^*B = C - P^*P + R$. Because $(P^*Pu, u) = (Pu, Pu) \geq 0$, we get that

$$\|Bu\|^2 = (B^*Bu, u) \leq C\|u\|^2 + (Ru, u)$$

and the L^2-boundedness of B follows from the L^2-boundedness of R.   □

Chapter 3

Symplectic Differential Geometry

We start this chapter with a brief review of the differential geometry we will need.

3.1. Vector fields

Let v be a C^∞ vector field on a C^∞ manifold M, $\dim M = n$. A *solution curve* of v is a differentiable mapping $t \to m(t): I \to M$, I an open interval in $\mathbb{R}$, such that

$$(3.1.1) \qquad \frac{dm(t)}{dt} = v(m(t)), \qquad t \in I.$$

As is well known (see, for instance, Coddington and Levinson [18]) there exists for each $m \in M$ exactly one maximal solution $t \to \Phi^t(m)$, defined on an open interval I_m in $\mathbb{R}$, such that $0 \in I_m$, $\Phi^0(m) = m$. The maximality of the interval I_m is reflected by the property that if, for instance, $b = \sup I_m < \infty$, then there exists for each compact subset K of M an $\varepsilon > 0$ such that $\Phi^t(m) \notin K$ for $t \in [b - \varepsilon, b[$. The set

$$(3.1.2) \qquad U = \{(t, m) \in \mathbb{R} \times M; t \in I_m\}$$

is an open subset of $\mathbb{R} \times M$ and Φ is a C^∞ mapping: $U \to M$.

It follows that for each $t \in \mathbb{R}$ the set

$$(3.1.3) \qquad U_t = \{m \in M; (t, m) \in U\}$$

is open in M, and

$$(3.1.4) \qquad \Phi^t: m \to \Phi^t(m)$$

is a C^∞ mapping: $U_t \to M$. If $m \in U_t$, $\Phi^t(m) \in U_{t'}$, then $m \in U_{t+t'}$ and

$$(3.1.5) \qquad \Phi^{t+t'}(m) = \Phi^{t'}(\Phi^t(m)).$$

Applying this to $t' = -t$ and using that $\Phi^0 =$ identity on M, we obtain that Φ^t is a diffeomorphism: $U_t \to U_{-t}$, with inverse Φ^{-t}.

P. Buser, *Fourier Integral Operators*, Modern Birkhäuser Classics,
DOI 10.1007/978-0-8176-8108-1_4, © Springer Science+Business Media, LLC 2011

A collection of mappings Φ^t with these properties is called a *local one-parameter group of transformations in M*, a *local dynamical system in M*, or a *flow in M* (all these names are common). With these Φ^t's in mind a vector field on M is also called an *infinitesimal transformation of M*.

Let Φ be differentiable: $M \to N$, N some other manifold of the same dimension, and assume that

$$(3.1.6) \qquad D\Phi_m \text{ is bijective: } T_m(M) \to T_{\Phi(m)}(N) \qquad \text{for all } m \in M.$$

Then we can define the *pull-back* $\Phi^* w$ on M of a vector field w on N by

$$(3.1.7) \qquad (\Phi^* w)(m) = D\Phi_m^{-1}(w(\Phi(m))), \qquad m \in M.$$

If u is a vector field on M inducing the flow Φ^t, then the *Lie-derivative of the vector field v with respect to the vector field u*, also called the *Lie product of u and v*, is defined by

$$(3.1.8) \qquad [u, v] = \frac{d}{dt}(\Phi^t)^* v\big|_{t=0}.$$

The Lie-product is natural with respect to transformations $\Phi \colon M \to N$ with bijective differentials, that is,

$$(3.1.9) \qquad \Phi^*[u, v] = [\Phi^* u, \Phi^* v]$$

for all vector fields u, v in N. In $\mathbb{R}^n$ we have the formula

$$(3.1.10) \qquad [u, v](m) = Dv_m(u(m)) - Du_m(v(m))$$

(obtained by direct computation), which in view of (3.1.9) enables us to compute $[u, v]$ in local coordinates. From (3.1.10) we see that $[u, v]$ is bilinear and antisymmetric in u and v, and the fact that second-order differentials are symmetric bilinear forms leads to *Jacobi's identity*:

$$(3.1.11) \qquad [u, [v, w]] + [v, [w, u]] + [w, [u, v]] = 0.$$

In other words, the C^∞ vector fields on M form a vector space $\chi(M)$ which is a Lie-algebra for this product.

Note also that

$$(3.1.12) \qquad \frac{d}{dt}(\Phi^t)^* v = \frac{d}{dt'}(\Phi^{t+t'})^* v\big|_{t'=0} = (\Phi^t)^*[u, v].$$

So, for example, $[u, v] = 0$ is equivalent to $(\Phi^t)^* v = v$ for all t, that is, v is *invariant under the u-flow*.

A C^∞ *tangent system* S *of dimension* k in M is a mapping which assigns to each $m \in M$ a k-dimensional linear subspace S_m of $T_m(M)$, and in such a way that for each $m_0 \in M$ there is a neighborhood U of m_0 and C^∞ vector fields $v_1, \ldots, v_k$ on U such that

$$(3.1.13) \qquad v_1(m), \ldots, v_k(m) \text{ span } S_m \quad \text{for all } m \in U.$$

An equivalent characterization reads: S is a k-dimensional C^∞ vector subbundle of $T(M)$.

A k-dimensional C^∞ submanifold S of M is called an *integral manifold* for S if

$$(3.1.14) \qquad T_m(S) = S_m \quad \text{for all } m \in S.$$

The tangent system S is called *integrable* if each $m \in M$ is contained in some integral manifold.

Theorem 3.1.1 (Frobenius). *Suppose* S *is spanned by the* C^∞ *vector fields* $v_1, \ldots, v_k$. *Then* S *is integrable if and only if*

$$(3.1.15) \qquad [v_i, v_j](m) \in S_m \quad \text{for all } m \in M, \ i, j = 1, \ldots, k.$$

Moreover, if (3.1.15) *is satisfied then there exists for each* $m_0 \in M$ *a diffeomorphism* $\Psi: U \times V \to W$, *such that* $\Psi(0, m) = m$ *and* $\Psi^* v_i$ *is a linear combination of* $\partial/\partial t_1, \ldots, \partial/\partial t_k$, *for all* $i = 1, \ldots, k$. *Here* U *is a neighborhood of the origin in* $\mathbb{R}^k$, *W* *is a neighborhood of* m_0 *in* M, *V* *is a* C^∞ *submanifold of* M *such that* $m_0 \in V$, $T_{m_0}(M) = S_{m_0} \oplus T_{m_0}(V)$. *Points in* $U \times V$ *are denoted by* $(t_1, \ldots, t_k, m)$, $t_i \in \mathbb{R}$, $m \in V$.

Proof. For the necessity of (3.1.15) we remark that if S is an integral manifold for S then $v_i(m) \in T_m(S)$ for all $m \in S$, so the $v_i|_S$ can be regarded as vector fields in S. Because of the unicity of the flow induced by a given smooth vector field, the $v_i|_S$-flow on S is equal to the restriction of the v_i-flow in M to S. Hence $[v_i, v_j](m) = [v_i|_S, v_j|_S](m) \in T_m(S)$ for all $m \in S$.

For the other part of the theorem, suppose that (3.1.15) holds and let V be as above. Define

$$(3.1.16) \qquad \Psi(t_1, \ldots, t_k, m) = \Phi_k^{t_k} \circ \cdots \circ \Phi_1^{t_1}(m),$$

where we denoted the v_i-flow by Φ_i^t. Shrinking V if necessary, this defines a C^∞ mapping: $\Psi: U \times V \to M$, U a neighborhood of 0 in $\mathbb{R}^k$; it is also obvious that $\Psi(0, m) = m$. The image of $D\Psi_{(0, m_0)}$ contains $v_1(m_0), \ldots, v_k(m_0)$

and $T_{m_0}(V)$, hence $D\Psi_{(0,m_0)}$ is bijective. It follows by the implicit function theorem that we can choose U, V so small that Ψ is a diffeomorphism from $U \times V$ onto an open neighborhood W of m_0.

The $\Psi^* v_i$ are linear combinations of $\partial/\partial t_1, \ldots, \partial/\partial t_k$ because $(\Phi_i^t)^* v_j$ is a linear combination of $v_1, \ldots, v_k$ for all i, j, all t.

Indeed, let Φ^t be the flow induced by a smooth vector field u, assume that

$$(3.1.17) \qquad [u, v_i] = \sum_j \lambda_{ij} v_j$$

for a matrix $\Lambda = \lambda_{ij}$ of smooth functions, for $i, j = 1, \ldots, k$. Let $A(t) = \alpha_{ij}(t)$ be the uniquely determined matrix of C^∞ functions on M satisfying

$$(3.1.18) \qquad \frac{d}{dt} A(t) = \Lambda(t) \circ A(t), \qquad A(0) = I.$$

Here $\Lambda(t) = \lambda_{ij}(t)$ is defined by

$$(3.1.19) \qquad \lambda_{ij}(t) = \lambda_{ij} \circ \Phi^t.$$

Then

$$(3.1.20) \qquad (\Phi^t)^* v_i = \sum_j \alpha_{ij}(t) \cdot v_j \qquad \text{for all } i, j = 1, \ldots, k, \text{ all } t.$$

Indeed,

$$\frac{d}{dt}(\Phi^t)^* v_i = (\Phi^t)^*[u, v_i] = (\Phi^t)^* \sum \lambda_{ij} v_j = \sum_j (\lambda_{ij} \circ \Phi^t) \cdot (\Phi^t)^* v_j.$$

So the left- and the right-hand side of (3.1.20) satisfy the same ordinary differential equation (with respect to t) and are equal at $t = 0$, so they are equal for all t. □

Remarks. From the proof of the theorem it also follows that the integral manifolds through m_0 of a smooth tangent system are locally uniquely determined, namely, equal to the image of $t \to \Psi(t, m_0)$. Note also that if $[v_i, v_j] = 0$ then $\Psi^* v_i = \partial/\partial t_i$ for all $i = 1, \ldots, k$.

In the above we treated vector fields as infinitesimal transformations, but one can also regard these as a special sort of first-order linear partial differential operator.

If A is an algebra over the field K, then $D: A \to A$ is called a K-*derivation* if

(3.1.21) $\qquad\qquad\qquad\qquad D$ is K-linear,

and

(3.1.22) $\qquad\qquad D(f \cdot g) = Df \cdot g + f \cdot Dg \qquad$ for all $f, g \in A$.

Now if v is a C^∞ vector field on M, then

(3.1.23) $\qquad (D_v f)(m) = Df_m(v(m)) = \dfrac{d}{dt} f(\Phi^t(m))\big|_{t=0}$

defines an $\mathbb{R}$-derivation of the algebra $C^\infty_{\mathbb{R}}(M)$ of real-valued C^∞ functions on M. Conversely it is easy to show that any derivation: $C^\infty_{\mathbb{R}}(M) \to C^\infty_{\mathbb{R}}(M)$ is equal to D_v for a unique C^∞ vector field v on M.

If D_1, D_2 are derivations of an algebra A, then the commutator $[D_1, D_2] = D_1 \circ D_2 - D_2 \circ D_1$ also is a derivation. One also verifies easily that the commutator product is antisymmetric and satisfies the Jacobi identity, that is the derivations form a Lie algebra, called $\mathrm{Der}(A)$, for the commutator product.

For C^∞ vector fields v_1, v_2 on M we have now

(3.1.24) $\qquad\qquad\qquad [D_{v_1}, D_{v_2}] = D_{[v_1, v_2]},$

as can be proved for instance by applying both sides to some $f \in C^\infty(M)$ and using (3.1.10) in local coordinates.

This leads to another proof of the assertion that the vector fields form a Lie-algebra with respect to the Lie product, in fact (3.1.24) expresses that $v \to D_v$ is a Lie-algebra isomorphism of $\chi(M)$ with the Lie-algebra $\mathrm{Der}(C^\infty_{\mathbb{R}}(M))$.

In local coordinates the identification of vector fields and derivations means that the vector field $(v_1(x), \dots, v_n(x))$ is identified with the first-order linear partial differential operator

$$D_v = \sum_{i=1}^{n} v_i(x) \cdot \frac{\partial}{\partial x_i}.$$

3.2. Differential forms

We now turn to the theory of differential forms of E. Cartan [14]. Let E be an n-dimensional vector space over $\mathbb{R}$. Denote by $\Lambda^k E^*$ the space of antisymmetric k-linear mappings

$$\omega: (e_1, \ldots, e_k) \mapsto \omega(e_1, \ldots, e_k): E^k \to \mathbb{R}.$$

For $k = 0$ we write $\Lambda^0 E^* = \mathbb{R}$, for $k = 1$ we have $\Lambda^1 E^* = E^* =$ the dual space of E.

For $\alpha \in \Lambda^k E^*$, $\beta \in \Lambda^\ell E^*$ the *exterior product* $\alpha \wedge \beta \in \Lambda^{k+\ell} E^*$ is defined by

$$(3.2.1) \qquad (\alpha \wedge \beta)(e_1, \ldots, e_{k+\ell}) = \sum_p \operatorname{sgn} P$$
$$\cdot \alpha(e_{p(1)}, \ldots, e_{p(k)}) \cdot \beta(e_{p(k+1)}, \ldots, e_{p(k+\ell)}),$$

where in the summation we take for each partition of $\{1, \ldots, k + \ell\}$ into subsets A, B, $\#(A) = k$, $\#(B) = \ell$, one permutation p such that $p(i) \in A$ for $i = 1, \ldots, k$, $p(i) \in B$ for $i = k + 1, \ldots, k + \ell$. Note that

$$\operatorname{sgn} p \cdot \alpha\big(e_{p(1)}, \ldots, e_{p(k)}\big) \cdot \beta\big(e_{p(k+1)}, \ldots, e_{p(k+\ell)}\big)$$
$$= \operatorname{sgn} q \cdot \alpha\big(e_{q(1)}, \ldots, e_{q(k)}\big) \cdot \beta\big(e_{q(k+1)}, \ldots, e_{q(k+\ell)}\big)$$

if $\{p(1), \ldots, p(k)\} = \{q(1), \ldots, q(k)\}$, $\{p(k + 1, \ldots, p(k + \ell)\} = \{q(k + 1), \ldots, q(k + \ell)\}$ so it does not matter which permutation we choose.

Proposition 3.2.1.

$(3.2.2) \quad \alpha \wedge (\beta \wedge \gamma) = (\alpha \wedge \beta) \wedge \gamma$ if $\alpha \in \Lambda^k E^*$, $\beta \in \Lambda^\ell E^*$, $\gamma \in \Lambda^m E^*$,

$(3.2.3)$ *If* $\alpha^1, \ldots, \alpha^n$ *form a basis of* E^* *then the* $\alpha^{i_1} \wedge \cdots \wedge \alpha^{i_k}$, $i_1 < i_2 < \cdots < i_k$ *form a basis of* $\Lambda^k E^*$. *In particular,* $\dim \Lambda^k E^* = \binom{n}{k}$.

$(3.2.4) \qquad \alpha \wedge \beta = (-1)^{k\ell} \beta \wedge \alpha$ if $\alpha \in \Lambda^k E^*$, $\beta \in \Lambda^\ell E^*$.

Proof. (3.2.2) is proved by showing that both sides are equal to

$$\sum \operatorname{sgn} p \cdot \alpha\big(e_{p(1)}, \ldots, e_{p(k)}\big) \cdot \beta\big(e_{p(k+1)}, \ldots, e_{p(k+\ell)}\big)$$
$$\cdot \gamma\big(e_{p(k+\ell+1)}, \ldots, e_{p(k+\ell+m)}\big),$$

where in the summation we take for each partition of $\{1, \ldots, k + \ell + m\}$ into subsets A, B, C, $\#(A) = k$, $\#(B) = \ell$, $\#(C) = \ell$ one permutation p

such that $p(i) \in A$, $i = 1, \ldots, k$, $p(i) \in B$, $j = k+1, \ldots, k+\ell$, $p(i) \in C$, $i = k+\ell+1, \ldots, k+\ell+m$.

For (3.2.3), write $\mathcal{I} =$ set of multi-indices $I = (i_1, \ldots, i_k)$ such that $1 \leq i_1 < i_1 < \cdots < i_k \leq n$. Write $\alpha^I = \alpha^{i_1} \wedge \cdots \wedge \alpha^{i_k}$, $e_I = (e_{i_1}, \ldots, e_{i_k}) \in E^k$, where $e_1, \ldots, e_n$ is a dual basis in E. Then $\alpha^I(e_{\mathcal{I}}) = 1$ if $I = J$ and $= 0$ if $I \neq J$. This proves that the α^I, $I \in \mathcal{I}$ are linearly independent. On the other hand,

$$\omega = \sum_{I \in \mathcal{I}} \omega(e_I) \cdot \alpha^I$$

because the left- and right-hand side have the same values on the e_J, $J \in \mathcal{I}$, hence on all of E^k.

Equation (3.2.4) follows from (3.2.3) and $\alpha \wedge \beta = -\beta \wedge \alpha$ if α, β are 1-forms.

If M is a C^∞ manifold then the $\Lambda^k T_x(M)^*$, $x \in M$ together form a C^∞ vector bundle $\Lambda^k T^* M$ over M in a natural way. Note that $\Lambda^1 T^* M = T^* M$ $=$ cotangent bundle of M, $\Lambda^0 T^* M \equiv M \times \mathbb{R}$.

A C^∞ k-form ω on M now is defined as a C^∞ section: $M \to \Lambda^k T^* M$, the space of C^∞ k-forms on M is denoted by $\Omega^k M$. Note that $\Omega^0 M = C^\infty_\mathbb{R}(M) =$ space of real-valued C^∞ functions on M. If $f \in C^\infty(M)$ then the total differential $Df: m \mapsto Df_m$ defines a 1-form on M. If $\alpha \in \Omega^k(M)$, $\beta \in \Omega^\ell(M)$ then $\alpha \wedge \beta \in \Omega^{k+\ell}(M)$ is defined by

(3.2.5) $\qquad (\alpha \wedge \beta)_m = \alpha_m \wedge \beta_m \qquad$ for all $m \in M$.

Now the *exterior differential* $d: \Omega^k(M) \to \Omega^{k+1}(M)$ is defined by

Theorem 3.2.2. *There is a unique family of mappings $d_k^U: \Omega^k(U) \to \Omega^{k+1}(U)$, $k = 0, 1, 2, \ldots$, U open in M, which all are denoted by d, such that*

(3.2.6) $\qquad d = D$ on $\Omega^0 = C^\infty$

(3.2.7) $\qquad d(f \cdot df_1 \wedge \cdots \wedge df_k) = df \wedge df_1 \wedge \cdots \wedge df_k$

for all $f, f_1, \ldots, f_k \in C^\infty$, d is additive;

(3.2.8) $\quad$ *d is a local operator in the sense that $d_k^V(\omega|_V) = d_k^U \omega|_V$ if U, V open in M, $V \subset U$, $\omega \in \Omega^k(U)$.*

Moreover this operator d has the following properties:

(3.2.9) $\quad d(\alpha \wedge \beta) = d\alpha \wedge \beta + (-1)^k \alpha \wedge d\beta$, $\alpha \in \Omega^k(U)$, $\beta \in \Omega^\ell(U)$,

(3.2.10) $\qquad d \circ d = 0$, *that is, $d_{k+1}^U \circ d_k^U = 0 \qquad$ for all k, U.*

For U open in $\mathbb{R}^n$, d is given by

$$d\omega_m(e_1, \ldots, e_{k+1})$$

(3.2.11)
$$= \sum_{i=1}^{k+1} (-1)^{i-1} D(\omega(e_1, \ldots, \hat{e}_i, \ldots, e_{k+1}))_m \cdot (e_i).$$

Here $\omega \in \Omega^k(U)$, $e_1, \ldots, e_{k+1}$ are constant vectors. $\hat{e}_i$ means that e_i should be deleted in the row $(e_1, \ldots, e_{k+1})$.

Proof. Let U be a coordinate neighborhood in M, $x_1, \ldots, x_n$ coordinate functions on U. Then the $Dx_{1(m)}, \ldots, Dx_{n(m)}$ are linearly independent in $T_m(M)^*$ for all $m \in U$, so each $\alpha \in \Omega^k(U)$ has a unique representation

$$\sum_{I \in J} c_I \cdot Dx_{i_1} \wedge \cdots \wedge Dx_{i_k}, \quad i = (i_1, \ldots, i_k),$$

according to (3.2.3), with some C^∞ functions c_I on U. So d is uniquely defined on the $\Omega^k(U)$ by the rules (3.2.6), (3.2.7), with $f_j = x_{i_j}$. Moreover, (3.2.8) holds with $V \subset U$.

Equation (3.2.9) follows from the derivation rule for D on $C^\infty(U)$. For (3.2.10) we remark that any $f \in C^\infty(U)$ is of the form $\tilde{f}(x_1, \ldots, x_n)$ for some $\tilde{f} \in C^\infty(\tilde{U})$, $\tilde{U}$ = image in $\mathbb{R}^n$ of the coordinate mapping: $m \mapsto (x_n(m), \ldots, x_n(m))$. The chain rule gives

$$Df = \sum \frac{\partial \tilde{f}(x_1, \ldots, x_n)}{\partial x_j} \cdot Dx_j,$$

so

$$d^2 f = \sum_{i,j} \frac{\partial^2 f(x_1, \ldots, x_n)}{\partial x_i \partial x_j} \cdot dx_i \wedge dx_j$$

which, however, vanishes because $\frac{\partial^2 f}{\partial x_i \partial x_j} = \frac{\partial^2 f}{\partial x_j \partial x_i}$ and $dx_i \wedge dx_j = -dx_j \wedge dx_i$. So we have proved (3.2.10) for $k = 0$, which implies it for arbitrary k using (3.2.9). This also implies (3.2.7) for arbitrary functions f_j.

Now the uniqueness of the d_k^U on coordinate neighborhoods gives that $d_k^U\big|_{\Omega^k(U \cap V)} = d_k^V\big|_{\Omega^k(U \cap V)}$ for all coordinate neighborhoods. Together with the locality of the operators d_k on coordinate neighborhoods we obtain that d is well-defined on M by $d\alpha\big|_U = d(\alpha\big|_U)$ for any coordinate neighborhood U in M.

Finally (3.2.11) is another way of stating (3.2.7) for coordinate functions f_j. $\square$

Is Φ a differentiable mapping from M to another smooth manifold N (not necessarily with bijective differential) then the *pull-back* to M of $\alpha \in \Omega^k(N)$ under Φ is defined by

$$(3.2.12) \qquad (\Phi^*\alpha)_m(e_1,\ldots,e_k) = \alpha_{\Phi(m)}(D\Phi_m e_1,\ldots,D\Phi_m e_k),$$
$$m \in M, \; e_1,\ldots,e_k \in T_m(M).$$

Proposition 3.2.3.

$$(3.2.13) \qquad \Phi^*(\alpha \wedge \beta) = \Phi^*\alpha \wedge \Phi^*\beta \qquad \text{for all } \alpha \in \Omega^k(N), \; \beta \in \Omega^\ell(N)$$

$$(3.2.14) \qquad \Phi^* d\alpha = d\Phi^*\alpha \qquad \text{for all } \alpha \in \Omega^k(N).$$

Proof. (3.2.13) is immediate. For (3.2.14) we only need to investigate the case $\alpha = f \cdot df_1 \wedge \cdots \wedge df_k$, where it reduces to $\Phi^* Df = D\Phi^* f$ for all $C^\infty(M)$. But this is the chain rule for the differential of the composition of functions. $\qquad \square$

If $v \in \chi(M)$, $\Phi^t = $ the v-flow in M then the *Lie derivative of* $\omega \in \Omega^k(M)$ *with respect to* v is defined by

$$(3.2.15) \qquad \mathcal{L}_v\omega = \frac{d}{dt}(\Phi^t)^*\omega\Big|_{t=0} \in \Omega^k(M).$$

The *inner product* $\omega \mid v \in \Omega^{k-1}(M)$ is defined by

$$(3.2.16) \qquad (\omega \mid v)_m(e_1,\ldots,e_{k-1}) = \omega_m(v_m,e_1,\ldots,e_{k-1})$$

for all $m \in M$, $e_1,\ldots,e_{k-1} \in T_m(M)$. Note that $\mathcal{L}_v f = Df \mid v = D_v f$ if $f \in C^\infty(M)$.

Proposition 3.2.4. *If Φ is $C^\infty \colon M \to N$ with bijective differentials, then*

$$(3.2.18) \qquad \Phi^*(\mathcal{L}_v\omega) = \mathcal{L}_{\Phi \cdot v}\Phi^*\omega$$

and

$$(3.2.19) \qquad \Phi^*(\omega \mid v) = (\Phi^*\omega \mid \Phi^*v)$$

for all $\omega \in \Omega^k(M)$, all k, and all $v \in \chi(M)$.

The proof is immediate. More substantial is:

Theorem 3.2.5. *If* $\omega \in \Omega^k(M)$, $v \in \chi(M)$, *then*

$$(3.2.20) \qquad \mathcal{L}_v\omega = d(\omega \mid v) + d(\omega) \mid v.$$

Proof. By induction with respect to k. For $k = 0$ it is trivial. Now suppose it is proven for $k - 1$. If $\omega = d\alpha$, $\alpha \in \Omega^{k-1}(M)$ then $\mathcal{L}_v d\alpha = d\mathcal{L}_v\alpha = d[d(\alpha \mid v) + \omega \mid v] = d(\omega \mid v)$ proves (3.2.20) for ω. However, $\mathcal{L}_v f\omega = vf \cdot \omega + fd(\omega \mid v) = d(f\omega \mid v) + d(f\omega) \mid v$ (here we used $(df \wedge \omega) \mid v = (vf) \cdot \omega - df \wedge (\omega \mid v)$), which proves (3.2.20) for all $f\omega$, $f \in C^\infty(M)$. But then it follows for all $\omega \in \Omega^k(M)$, which are locally finite sums of elements of the form

$$f \, df_1 \wedge df_k = f \, d(f_1 \cdot df_2 \wedge \cdots \wedge df_k). \qquad \qquad \square$$

Theorem 3.2.6. (Poincaré lemma). *For each* $m_0 \in M$ *there is a neighborhood* U *of* m_0 *such that for each* $\omega \in \Omega^k(U)$, $k \geq 1$:

$$(3.2.21) \qquad d\omega = 0 \Leftrightarrow \exists \alpha \in \Omega^{k-1}(U) \ \text{such that} \ \omega = d\alpha.$$

Proof. Let Φ^t be the flow of a vector field v such that $v(m_0) = 0$, Re $\lambda > 0$ for the eigenvalues λ of $L = Dv_{m_0}$. This means that there is a neighborhood U of m_0 and constants $C, \varepsilon > 0$ such that $\lim\limits_{t \to -\infty} \Phi^t(m) = m_0$ and $\|D\Phi_m^t\| \leq ce^{\varepsilon t}$ for all $m \in U$, $t \leq 0$. Now define

$$(3.2.22) \qquad H\omega = \left(\int_{-\infty}^0 (\Phi^t)^*\omega \, dt \right) \mid v.$$

Then

$$dH\omega = \mathcal{L}_v \int_{-\infty}^0 (\Phi^t)^*\omega \, dt - \left(d \int_{-\infty}^0 (\Phi^t)^*\omega \, dt \right) \mid v$$

$$= (\Phi^t)^*\omega \Big|_{-\infty}^0 - \left(\int_{-\infty}^0 (\Phi^t)^*d\omega \, dt \right) \mid v = \omega - H \, d\omega,$$

that is,

$$(3.2.23) \qquad d \circ H + H \circ d = \text{identity on } \Omega^k(U).$$

This solves (3.2.21) with $\alpha = H\omega$. $\qquad \qquad \square$

If Ω^k denotes the sheaf of germs of C^∞ sections of $\Lambda^k T^* M$ then the Poincaré lemma says that

$$(3.2.24) \qquad 0 \to \mathbb{R} \hookrightarrow C^\infty = \Omega^0 \xrightarrow{d_0} \Omega^1 \xrightarrow{d_1} \Omega^2 \xrightarrow{d_2}$$

is an exact sequence of sheaves. Because the sheaves Ω^k are fine this implies:

Theorem 3.2.7 (de Rham [63], Ch. 4). *There is a natural isomorphism:*

$$(3.2.25) \qquad \operatorname{Ker} d_k^M / \operatorname{Range} d_{k-1}^M \cong H^k(M, \mathbb{R})$$

equal to the k-th Čech-cohomology group of M with values in $\mathbb{R}$. See, for instance, Gunning [37] for the sheaf theory used here.

For example, every $\omega \in \Omega^1(M)$ such that $d\omega = 0$ is the differential of a function if and only if $H^1(M, \mathbb{R}) = 0$. Also observe that (3.2.25) implies that $H^k(M, \mathbb{R}) = 0$ for $k > n$, because $\Omega^k = 0$ for $k > n$.

Remark. For a moment I was tempted to think that the name "exterior differential" came from its role in *Stokes' formula*

$$(3.2.26) \qquad \int_\gamma d\omega = \int_{\partial\gamma} \omega,$$

valid, for instance, for $\omega \in \Omega^k(M)$, γ = compact oriented $(k+1)$-dimensional manifold with smooth boundary $\partial\gamma$ (which automatically also is oriented). But this is historically not true: Grassmann [34] introduced exterior algebra without reference to Stokes' formula and d was called "exterior" because of its close relation to the exterior product $\wedge$.

The relation between d and $\wedge$ can be illustrated as follows. In $M = \mathbb{R}^n$, d is a differential operator with its *symbol* at (x, ξ) equal to the mapping

$$(3.2.27) \qquad \omega \mapsto i\xi \wedge \omega.$$

Proof.

$$d\omega_x = d(2\pi)^{-n} \iint e^{i\langle x-y, \xi\rangle} \omega_y \, dy \, d\xi$$

$$= (2\pi)^{-n} \iint d_x e^{i\langle x-y, \xi\rangle} \wedge \omega_y \, dy \, d\xi$$

$$= (2\pi)^{-n} \iint e^{i\langle x-y, \xi\rangle} i\xi \wedge \omega_y \, dy \, d\xi.$$

3.3. The canonical 1- and 2-form in $T^*(X)$

Let X be an n-dimensional C^∞ manifold. The *cotangent bundle* $T^*(X)$ is defined as the disjoint union of the $T_x(X)^* =$ dual space of the tangent space of X at x, $x \in X$. Points $\xi \in T_x(X)^*$ will be denoted by (x, ξ) when regarded as elements of $T^*(X)$. The projection $\pi: T^*(X) \to X$ assigns to each $\xi \in T_x(X)^*$ the base point x.

Let κ be a coordinatization of a coordinate neighborhood U in X. On $\pi^{-1}(U) \subset T^*(X)$ the *induced coordinatization* $\tilde{\kappa}$ is defined by:

$$(3.3.1) \qquad \tilde{\kappa}(x, \xi) = \left(\kappa(x), ({}^t D\kappa_x)^{-1}(\xi)\right) \in \mathbb{R}^n \times \mathbb{R}^n.$$

Note that $D\kappa_x$ is linear: $T_x(X) \to \mathbb{R}^n$ so ${}^t D\kappa_x$ is linear: $\mathbb{R}^n \cong (\mathbb{R}^n)^* \to T_x(X)^*$.

It is easily verified that the induced coordinatizations form a collection of local trivializations making $T^*(X)$ into a C^∞ vector bundle over X with the projection π.

If $(x, \xi) \in T^*(X)$ then $D\pi_{(x,\xi)}$ is linear: $T_{(x,\xi)}(T^*(X)) \to T_x(X)$. In turn ξ is linear: $T_x(X) \to \mathbb{R}$, this leads to:

Definition 3.3.1. The *canonical 1-form on* $T^*(X)$ is the 1-form α given by

$$(3.3.2) \qquad \alpha_{(x,\xi)} = \xi \circ D\pi_{(x,\xi)} \qquad \text{for all} \quad (x, \xi) \in T^*(X).$$

Proposition 3.3.1.

$$(3.3.3) \qquad \lambda = \lambda^* \alpha \quad \text{for every} \ \lambda \in \Omega^1(X),$$

which in the right-hand side is regarded as a section $\lambda: X \to T^*(X)$.

Proof. Using the chain rule, we get

$$(\lambda^* \alpha)_x = \alpha_{(x, \lambda_x)} \circ D\lambda_x = \lambda_x \circ D\pi_{(x, \lambda_x)} \circ D\lambda_x(u) = \lambda_x$$

because $\pi \circ \lambda = $ identity in X. $\qquad\qquad\qquad\qquad\qquad\qquad$ □

Definition 3.3.2. $\sigma = d\alpha$ is called the *canonical 2-form on* $T^*(X)$.

Proposition 3.3.2. *A C^∞ submanifold Λ of $T^*(X)$ is locally equal to the graph of $d\phi$, ϕ a C^∞ function on X, if and only if*

$$(3.3.4) \qquad \qquad \dim \Lambda = n,$$

(3.3.5) $$D\pi_{(x,\xi)}: T_{(x,\xi)}(\Lambda) \to T_x(X)$$

is injective for each $(x, \xi) \in \Lambda$, and finally

(3.3.6) $\sigma_{(x,\xi)}(u, v) = 0$ for all $u, v \in T_{(x,\xi)}(\Lambda)$, $(x, \xi) \in \Lambda$.

Proof. (3.3.4), (3.3.5) mean exactly that locally Λ is equal to the graph of a section $\lambda: X \to T^*(X)$ whereas (3.3.6) means that $\lambda^*\sigma = 0$. In view of (3.3.3) this means that $d\lambda = 0$, which in virtue of Poincaré's lemma is equivalent to $\lambda = d\phi$, locally, for some $\phi \in C^\infty$.   □

In induced coordinates $(x_1, \ldots, x_n, \xi_1, \ldots, \xi_n)$ we have

(3.3.7) $$\alpha = \sum_{j=1}^{n} \xi_j \, dx_j$$

and hence

(3.3.8) $$\sigma = \sum_{j=1}^{n} d\xi_j \wedge dx_j.$$

In other words, if $\begin{pmatrix} a \\ b \end{pmatrix}$, $\begin{pmatrix} a' \\ b' \end{pmatrix} \in \mathbb{R}^n \times (\mathbb{R}^n)^*$ (written as column vectors) then

(3.3.9) $$\sigma\left(\begin{pmatrix} a \\ b \end{pmatrix}, \begin{pmatrix} a' \\ b' \end{pmatrix}\right) = b(a') - b'(a).$$

In particular, we obtain

Proposition 3.3.3. σ is a nondegenerate 2-form on $M = T^*(X)$, that is,

(3.3.10)
If $m \in M$, $u \in T_m(M)$, $\sigma_m(u, v) = 0$ for all $v \in T_m(M)$,
then $u = 0$.

Proof. It suffices to prove it in induced coordinates and there $b(a') - b'(a) = 0$ for all a', b' implies $b = 0$ (taking $b' = 0$) and $a = 0$ (taking $a' = 0$).   □

Definition 3.3.3. An antisymmetric nondegenerate bilinear form on a finite dimensional vector space E is called a *symplectic form* on E. A *symplectic vector space* is a pair (E, σ) consisting of a finite-dimensional vector space E and a symplectic form σ on E.

A *symplectic form on a manifold* M is a 2-form σ on M such that

(3.3.11) $$d\sigma = 0,$$

and

(3.3.12) σ_m is a symplectic form on $T_m(M)$, for each $m \in M$.

A *symplectic manifold* is a pair (M, σ) consisting of a manifold M and a symplectic form σ on M.

It follows from the previous that for any manifold X, $T^*(X)$ is a symplectic manifold with the canonical 2-form on $T^*(X)$. Note that $d\sigma = 0$ follows from $\sigma = d\alpha$.

Remark. In view of (3.3.8) it would perhaps have been better to write $(\xi, x) \in T^*(X)$ if $\xi \in T_x(X)^*$, instead of (x, ξ). In fact this is the custom in the mechanics-oriented literature, with ξ replaced by p and x by q. It should be remarked here that in many texts the 2-form $\sum dx_j \wedge d\xi_j$ is used as the canonical form, leading to a change of sign in the formulas.

3.4. Symplectic vector spaces

Let (E, σ) be a symplectic vector space over a field k. The bilinear form σ on E induces a linear mapping: $E \to E^*$, which also will be denoted by σ, defined by

(3.4.1) $(\sigma e_1)(e_2) = \sigma(e_1, e_2)$ for all $e_1, e_2 \in E$.

The nondegeneracy of the bilinear form σ is equivalent to the injectivity, hence bijectivity of the mapping $\sigma \colon E \to E^*$. (Note that dim $E^* =$ dim E.)

If L is a linear subspace of E then we define its *orthocomplement* L^σ *in* E *with respect to* σ by

(3.4.2) $$L^\sigma = \{e \in E;\ \sigma(e, \ell) = 0 \quad \text{for all } \ell \in L\}.$$

As for the orthocomplement with respect to an inner product, we have the following rules.

(3.4.3) $$L \subset M \Rightarrow M^\sigma \subset L^\sigma.$$

(3.4.4) $$(L^\sigma)^\sigma = L$$

(3.4.5) $$(L \cap M)^\sigma = L^\sigma + M^\sigma, \qquad (L + M)^\sigma = L^\sigma \cap M^\sigma$$

(3.4.6) $\dim L^\sigma = \dim E - \dim L.$

However the antisymmetry of σ also has some rather peculiar consequences. Because $\sigma(e,e) = 0$ for each $e \in E$, we have $L \subset L^\sigma$ for each one-dimensional subspace L of E. More generally a linear subspace L of E is called *isotropic with respect to* σ, if $L \subset L^\sigma$, that is, $\sigma(e_1, e_2) = 0$ for all $e_1, e_2 \in L$. (Example: the condition (3.3.6) in Proposition 3.3.2 says that $T_{(x,\xi)}(\Lambda)$ is isotropic for all $(x,\xi) \in \Lambda$.) A maximally isotropic subspace of E is called *Lagrangian* in (E, σ).

Proposition 3.4.1. *Let (E, σ) be a symplectic vector space. A linear subspace L of E is Lagrangian in (E, σ) if and only if $L = L^\sigma$. The dimension of E is even, say $= 2n$. An isotropic linear subspace L of E is Lagrangian if and only if $\dim L = n$.*

Proof. If $L \subsetneq L^\sigma$ then we can choose $e \in L^\sigma$, $e \notin L$. It follows that $\sigma(\ell_1 + \alpha_1 e, \ell_2 + \alpha_2 e) = 0$ for all $\ell_1, \ell_2 \in L$, $\alpha_1, \alpha_2 \in k$. So $L + k \cdot e$ is isotropic, $L \subsetneq L + k \cdot e$. This proves the first assertion. Now let L be Lagrangian (the existence of such L is trivial). Then $\dim L = \dim L^\sigma = \dim E - \dim L$, hence $\dim E = 2 \cdot \dim L$. Conversely $L \subset L^\sigma$, $\dim E = 2 \cdot \dim L$ implies that $L \subset L^\sigma$, $\dim L = \dim L^\sigma$, hence $L = L^\sigma$. $\qquad\square$

Definition 3.4.1. Let (E, σ), (F, τ) be symplectic vector spaces. A *symplectic mapping* $A: (E, \sigma) \to (F, \tau)$ is a linear mapping $A: E \to F$ such that $A^* \tau = \sigma$, that is, $\sigma(e_1, e_2) = \tau(Ae_1, Ae_2)$ for all $e_1, e_2 \in E$. A symplectic mapping is automatically injective. If $\dim E = \dim F$ then A is called a *symplectic isomorphism*.

Theorem 3.4.2. *For each Lagrangian subspace L there exists a Lagrangian subspace M such that $E = L \oplus M$. The mapping $A: \ell \oplus m \to (\ell, \sigma m|_L)$ is a symplectic isomorphism: $(E, \sigma) \to L \times L^*$, where the latter space is provided with the canonical symplectic form*

$$\left(\begin{pmatrix} a \\ b \end{pmatrix}, \begin{pmatrix} a' \\ b' \end{pmatrix} \right) \to b(a') - b'(a).$$

Proof. Let $M \cap L = (0)$, $M \subsetneq M^\sigma$. Then $M^\sigma \not\subset M + L$, because $M^\sigma \subset M + L$ implies $M \supset M^\sigma \cap L^\sigma = M^\sigma \cap L$, hence $M^\sigma \cap L = (0)$. But $\dim M^\sigma > n$, $\dim L = n$ makes this impossible. Choose $e \in M^\sigma$,

$e \notin M + L$. Then $M + [e]$ is isotropic and $(M + k \cdot e) \cap L = (0)$. This leads by induction to a Lagrangian subspace M such that $M \cap L = (0)$.

For the second part we remark that $\sigma(\ell + m, \ell' + m') = \sigma(\ell, m') + \sigma(m, \ell') = (\sigma m)(\ell') - (\sigma m')(\ell)$ if $\ell, \ell' \in L$, $m, m' \in M$. So A is symplectic and an isomorphism because $\dim E = 2 \dim L = \dim (L \times L^*)$.   □

If L_1, L_2 are vector spaces over k of the same dimension n, then a mapping $A: L_1 \times L_1^* \to L_2 \times L_2^*$ is called a *linear canonical transformation* if it is symplectic for the canonical symplectic forms in $L_1 \times L_1^*$ and $L_2 \times L_2^*$, respectively. If C is a linear map: $L_1 \to L_2$ then the mapping

$$\tilde{C}: \begin{pmatrix} a \\ b \end{pmatrix} \to \begin{pmatrix} Ca \\ ({}^tC)^{-1}b \end{pmatrix}$$

is a canonical transformation: $L_1 \times L_1^* \to L_2 \times L_2^*$, called the *canonical transformation induced by C*.

So every linear coordinatization $C: L \to k^n$ induces a linear canonical transformation $\tilde{C}: L \times \tilde{L} \to k^n \times (k^n)^*$, and Theorem 3.4.2 expresses that every symplectic vector space is symplectically isomorphic to $k^n \times (k^n)^*$ provided with the canonical symplectic form, for some n. Note that the bijective linear mapping $A: E \to k^n \times (k^n)^*$ sending the vectors $(e_1, \ldots, e_n, f_1, \ldots, f_n)$ into the standard basis of $k^n \times (k^n)^*$, is a symplectic transformation if and only if for all $i, j = 1, \ldots, n$:

(3.4.7)
$$\sigma(e_i, e_j) = 0$$
$$\sigma(f_i, f_j) = 0$$
$$\sigma(f_i, e_j) = 1 \text{ when } i = j, \text{ and } = 0 \text{ when } i \neq j.$$

Such a mapping is called a *canonical coordinatization* of (E, σ) and $(e_1, \ldots, e_n, f_1, \ldots, f_n)$ is called a *canonical basis* of (E, σ).

In the rest of this section we collect some more properties of symplectic automorphisms and of the manifold of all Lagrangian subspaces in (E, σ). Because we will not use this in the rest of this chapter, this can be skipped at first reading. We start by investigating Jordan normal forms of symplectic automorphisms.

Theorem 3.4.3. *Let (E, σ) be a symplectic vector space over k and let A be a symplectic mapping: $(E, \sigma) \to (E, \sigma)$ such that all its eigenvalues are contained in k.*

*Then there exists a Lagrangian subspace L of (E,σ) such that $A(L) \subset$
L. On each canonical basis $e_1,\ldots,e_n$, $f_1,\ldots,f_n$ such that $e_1,\ldots,e_n \in L$,
the mapping A has a matrix of the form*

$$(3.4.8) \qquad \begin{pmatrix} \alpha & \varepsilon\alpha \\ 0 & ({}^t\alpha)^{-1} \end{pmatrix}, \quad \text{for some symmetric matrix } \varepsilon.$$

*Conversely, each such matrix (with α nonsingular) on a canonical basis
represents a symplectic automorphism.*

Proof. Let L be invariant under A and $L \subsetneq L^\sigma$. Because A is symplectic,
L^σ is also invariant under A. Define the mapping $A_{L^\sigma/L} \colon L^\sigma/L \to L^\sigma/L$ by

$$(3.4.9) \qquad A_{L^\sigma/L}(e + L) = Ae + L.$$

Let λ be an eigenvalue of $A_{L^\sigma/L}$. Then λ is also an eigenvalue of A, hence
$\lambda \in k$ and we can find $f \in L^\sigma/L$, $f \neq 0$ such that $Af = \lambda f$. $f = e + L$
for some $e \in L^\sigma$, $e \notin L$. It follows that $k \cdot e + L$ is isotropic and invariant
under A. Repeating this procedure we finally end up with an invariant
Lagrangian subspace. The rest follows by a direct computation. $\qquad\square$

By choosing for $e_1,\ldots,e_n$ a basis of L on which L has Jordan form, we
obtain a clear picture of the possibilities for the Jordan forms of symplectic
automorphisms.

Corollary 3.4.4. *Suppose now that $k = \mathbb{R}$. Then the eigenvalues λ, $\bar\lambda$,
λ^{-1}, and $(\bar\lambda)^{-1}$ of A all have the same multiplicities. The multiplicities of
$+1$ and -1 are even.*

Proof. Complexifying (E,σ) we can apply Theorem 3.4.3 and from (3.4.8)
we see that λ^{-1} has the same multiplicity as λ. $\bar\lambda$ has the same multiplicity
as λ because A is real and because of the previous $(\bar\lambda)^{-1}$ also has this
multiplicity. In the case $\lambda = \lambda^{-1}$, that is $\lambda = \pm 1$, we see again from (3.4.8)
that the multiplicities are even. $\qquad\square$

If $k = \mathbb{R}$ or $\mathbb{C}$ the symplectic automorphisms form a closed subgroup
$\mathrm{Sp}(E,k)$ of $\mathrm{GL}(E,k)$ and therefore a Lie group; it is in fact a classical
one. The Lie algebra $\mathrm{sp}(E,k)$ of $\mathrm{Sp}(E,k)$, that is, the tangent space of
$\mathrm{Sp}(E,k)$ at the identity, consists of the linear mappings $A \colon E \to E$ such that
$\sigma(Ae_1,e_2) + \sigma(e_1,Ae_2) = 0$. Such mappings are also called *infinitesimal
linear symplectic transformations*. As in Theorem 3.4.3 one proves:

Theorem 3.4.5. *Let A be an infinitesimal linear symplectic transformation, assume that all eigenvalues of A are in k. Then there is a Lagrangian subspace L of E that is invariant under A. On each canonical basis $e_1, \ldots, e_n, f_1, \ldots, f_n$ for which $e_1, \ldots, e_n$ is a basis of L, A has a matrix of the form*

$$(3.4.10) \qquad \begin{pmatrix} \alpha & \varepsilon \\ 0 & -{}^t\alpha \end{pmatrix} \quad \text{with symmetric } \varepsilon.$$

Conversely every matrix of the form (3.4.10) *represents an infinitesimal symplectic transformation.*

Corollary 3.4.6. *Let A be a real infinitesimal linear symplectic transformation. Then the eigenvalues λ, $-\lambda$, $\bar{\lambda}$ and $-\bar{\lambda}$ all have the same multiplicity. The multiplicity of the eigenvalue 0 is even.*

We now turn to the study of the collection $\Lambda(E)$ of all Lagrangian subspaces of a real symplectic vector space (E, σ) of dimension $2n$. Most results are due to Arnol'd [5] and Hörmander [40], Section 3.3.

Theorem 3.4.7. $\Lambda(E)$ *is a connected regular algebraic subvariety of the Grassmann variety $G_{E,n}$ of n-dimensional linear subspaces of E, dim $\Lambda(E)$ $= \frac{1}{2}n(n+1)$. For any $M \in \Lambda(E)$, $\Lambda^0(E, M) = \{L \in \Lambda(E); L \cap M = (0)\}$ is (algebraically) diffeomorphic to the vector space Symm(L) of all symmetric bilinear forms on L, for any $L \in \Lambda^0(E, M)$. Moreover, $\Lambda^0(E, M)$ is open and dense in $\Lambda(E)$. Finally $T_L(\Lambda(E))$ is canonically isomorphic to Symm(L) for every $L \in \Lambda(E)$.*

Proof. Let $L, M \in \Lambda(E)$, $L \cap M = (0)$, these exist according to Theorem 3.4.2. Then each n-dimensional subspace L' that is transversal to M is of the form $\{x + Ax; x \in L\}$ for some linear map $A: L \to M$. Then

$$(3.4.11) \qquad Q(L')(x, y) = \sigma(Ax, y), \qquad x, y \in L$$

defines a bilinear form on L, which is symmetric if and only if L' is Lagrangian. So Q defines a bijective mapping: $\Lambda^0(E, M) \to$ Symm(L), the inverse being an algebraic embedding: Symm$(L) \to G_{E,n}$. So the Q's form a set of local coordinatizations of $\Lambda(E)$; one can prove that 2^n of the $\Lambda^0(E, N)$ cover $\Lambda(E)$. This proves the first part of the theorem.

Clearly $\Lambda^0(E, M)$ is open in $\Lambda(E)$. For its density we choose, for any $L \in \Lambda(E)$, an $M' \in \Lambda^0(E, M) \cap \Lambda^0(E, L)$. See the proof of Theorem 3.4.2; at each step one now has to take e in the complement of the union of two linear suspaces. Then $\Lambda^0(E, M')$ is an open neighborhood of L, identified with Symm(M), in which $\Lambda^0(E, M')$ corresponds to the nonsingular elements of Symm(M), which lie dense.

Clearly Q depends on the choice of L, M, but for fixed $L \in \Lambda(E)$ its differential at L does not depend on M. Proof: Let $\tilde{M} \in \Lambda^0(E, L)$, then $\tilde{M} = \{z + Bz; z \in M\}$ for some linear map $B: M \to L$. If L' is close enough to L, L' still is transversal both to M and $\tilde{M}$ and we can write $L' = \{x + Ax; x \in L\} = \{y + \tilde{A}y; y \in L\}$, for some linear maps $A: L \to N$, $\tilde{A}: L \to \tilde{N}$.

Then $x + Ax = y + \tilde{A}y$, $\tilde{A}y = z + Bz$ for some $z \in N$, so $z = Ax$, $y = x - Bz$, $x + Ax = (x - BAx) + \tilde{A}(x - BAx)$. Taking symplectic product with $u \in L$:

$$(3.4.12) \qquad \sigma(Ax, u) = \sigma(\tilde{A}x) - \sigma(\tilde{A}BAx, u).$$

The second term in the right-hand side vanishes of second order when $L' \to L$, so Q and $\tilde{Q}$ have the same differential (= first-order approximation) at $L' = L$. $\qquad \square$

Theorem 3.4.8. *For any $k \geq 1$, $M \in \Lambda(E)$ the set*

$$(3.4.13) \qquad \Lambda^k(E, M) = \{L \in \Lambda(E); \; \dim L \cap M = k\}$$

is the regular part of its closure $\overline{\Lambda^k(E, M)} = \bigcup_{\ell \geq k} \Lambda^\ell(E, M)$, which in turn is a connected algebraic subvariety of $\Lambda(E)$ of codimension $\frac{1}{2}k(k+1)$ in $\Lambda(E)$. If $L \in \Lambda^k(E, M)$, then $T_L(\Lambda^k(E, M))$ corresponds to the $Q \in$ Symm(L) such that $Q(x, y) = 0$ for all $x, y \in L \cap M$.

Proof. In view of Theorem 3.4.7 also $\Lambda^0(E, L) \cap \Lambda^0(E, M)$ is dense in $\Lambda(E)$ so it contains at least some N. Let Q be the coordinization: $\Lambda^0(E, N) \to$ Symm(L) described in the proof of Theorem 3.4.7. Then, if $L \in \Lambda^k(E, M)$, we have $L' \in \Lambda^k(E, M) \cap \Lambda^0(E, N)$ if and only if $\dim \ker(Q(M) - Q(L')) = k$. Note that $Q(L) = 0$.

On a suitable basis of L we can write

$$Q(M) = \begin{pmatrix} 0 & 0 \\ 0 & A \end{pmatrix}, \qquad Q(L') = \begin{pmatrix} B & C \\ {}^t C & D \end{pmatrix},$$

where A is a nonsingular $(n - k) \times (n - k)$ matrix.

If D is sufficiently small, then $A - D$ is still nonsingular and rank $(Q(M) - Q(L')) = n - k$ if and only if $B = -CS$, ${}^t C = (A - D)S$ for some

S. So the equation $L' \in \Lambda^k(E, M)$ reads $B = -C(A - D)^{-1}\,{}^tC$ if L' is close to L. This proves that $\Lambda^k(E, M)$ is a smooth algebraic manifold of codimension $\frac{1}{2}k(k + 1)$ and also that $T_L(\Lambda^k(E, M))$ corresponds to the symmetric matrices $\begin{pmatrix} B & C \\ {}^tC & D \end{pmatrix}$ such that $B = 0$.

To prove that $\Lambda^k(E, M)$ is connected we remark that $L \to L \cap M$ defines a fibration: $\Lambda^k(E, M) \to G_{M,k}$, with fiber over M_0 equal to $\Lambda(M_0^\sigma / M_0)$. To explain this, let $M_0 \subset M$, $\dim M_0 = k$. Then $M_0 \subset M = M^\sigma \subset M_0^\sigma$. Furthermore M_0^σ / M_0 is a symplectic vector space with symplectic form $\sigma_{\text{mod } M_0}$ defined by

$$(3.4.14) \qquad \sigma_{\text{mod } M_0}(e_1 + M_0, e_2 + M_0) = \sigma(e_1, e_2), \qquad e_1, e_2 \in M_0^\sigma.$$

If $L \in \Lambda(E)$, $L \cap M = M_0$, then also $M_0 \subset L = L^\sigma \subset M_0^\sigma$ and it is easily verified that $L \mapsto L/M_0$ is an isomorphism between the fiber over M_0 and $\Lambda(M_0^\sigma / M_0)$. Because both $G_{M,k}$ and $\Lambda(M_0^\sigma / M_0)$ are connected it follows that $\Lambda^k(E, M)$ is connected. $\qquad\square$

Theorem 3.4.9. $\Lambda(E) \setminus \Lambda^0(E, M)$ *defines an oriented cycle* μ *of codimension 1 in* $\Lambda(E)$. $\pi_1(\Lambda(E)) \simeq \mathbb{Z}$. *The mapping* $\alpha\colon \gamma \to \gamma \cdot \mu =$ *intersection number of* γ *and* μ, $\gamma \in \pi_1(\Lambda(E))$, *defines a generator of* $H^1(\Lambda(E), \mathbb{Z})$. α *is called the Maslov–Arnol'd index of* (E, σ).

Proof. Since only a variety of codimensions 3 in $\Lambda(E)$ is attached to its regular part $\Lambda^1(E, M)$, $\Lambda(E) \setminus \Lambda^0(E, M)$ defines a chain of codimension 1 in $\Lambda(E)$ without boundary. Moreover, it is oriented by defining as the positive side at $L \in \Lambda^1(E, M)$ the set of all $u \in T_L(\Lambda(E))$ such that the corresponding quadratic form on L is positive definite on the line $L \cap M$.

To prove that $\pi_1(\Lambda(E)) \simeq \mathbb{Z}$, and α is a generator of $M^1(\Lambda(E), \mathbb{Z})$ we prove that α is injective and exhibit an element γ_0 of $\pi_1(\Lambda(E))$ such that $\gamma_0 \cdot \mu = 1$.

Suppose $\gamma \in \pi_1(\Lambda(E))$, $\gamma \cdot \mu = 0$. Represent γ by a closed differentiable curve $\gamma(t)$, intersecting $\Lambda(E) \setminus \Lambda^0(E, M)$ only finitely often and only at $\Lambda^1(E, M)$ and transversally. Then

$$\gamma \cdot \mu = \sum_{\gamma(t) \in \Lambda^1(E, M)} \pm 1$$

with $+1$ if $\gamma(t)$ crosses $\Lambda^1(E, M)$ from $-$ to $+$ and with -1 otherwise. Now let $\gamma(t_1)$ be a positive and $\gamma(t_2)$ a subsequent negative crossing, that is, $\gamma(t) \in \Lambda^0(E, M)$ if $t_1 < t < t_2$. Connect $\gamma(t_1)$ and $\gamma(t_2)$ by a differentiable curve δ in $\Lambda^1(E, M)$, which can be accompanied by nearby curves δ_+, respectively, δ_- at the positive and negative side of $\Lambda^1(E, M)$, connecting $\gamma(t_1 + \varepsilon)$ with $\gamma(t_2 - \varepsilon)$ and $\gamma(t_1 - \varepsilon)$ with $\gamma(t_2 + \varepsilon)$, respectively. Here $\varepsilon > 0$ is small. Because $\Lambda^0(E, M)$ is simply connected, the curve $\gamma(t)$, $t_1 + \varepsilon \leq t \leq t_2 - \varepsilon$, followed by $(\delta_+)^{-1}$ is a contractible closed curve. A simple geometric argument shows that the curve $\gamma(t)$, $t_1 - \varepsilon \leq t \leq t_1 + \varepsilon$, followed by δ_+, then $\gamma(t)$, $t_2 - \varepsilon \leq t \leq t_2 + \varepsilon$ and then by $(\delta_-)^{-1}$ also is a contractible closed curve. So γ is homotopic to $\gamma(t)$, $t \leq t_1 - \varepsilon$, followed by $(\delta_-)^{-1}$, and then $\gamma(t)$, $t \geq t_2 + \varepsilon$. However this curve has two intersections less with $\Lambda^1(E, M)$. Eliminating in this way all intersections, we arrive at the conclusion that γ is homotopic to a point.

Now we construct γ_0. Let P be a nonisotropic two-dimensional subspace of (E, σ). Then (P, σ) is a symplectic vector space and $E = P \oplus P^\sigma$. On canonical coordinates in P, define

$$\gamma_P(\phi) = \left\{ \begin{pmatrix} x \cos \phi \\ x \sin \phi \end{pmatrix}; \ x \in \mathbb{R} \right\}, \quad \phi \text{ from } 0 \text{ to } \pi.$$

γ_P is a closed curve in $\Lambda(P)$ and intersects the ξ-axis only for $\phi = \pi/2$. At that point, its derivative $\gamma_P'(\frac{\pi}{2})$ is represented by the quadratic form

$$\xi \to \frac{d}{d\phi} \sigma \left(\begin{pmatrix} \xi \cos \phi / \sin \phi \\ 0 \end{pmatrix}, \begin{pmatrix} 0 \\ \xi \end{pmatrix} \right)_{\phi = \pi/2} = \xi^2,$$

so the intersection is positive. To lift this example to E, let L_0, M_0 be transversal Lagrangian subspaces of the symplectic vector space P^σ. Define

$$\gamma_0(\phi) = \gamma_P(\phi) + L_0, \qquad M = \gamma_P(\pi/2) + M_0.$$

Then $\gamma_0(\phi)$ intersects $\Lambda(E) \setminus \Lambda^0(E, M)$ only for $\phi = \pi/2$, in $\Lambda^1(E, M)$ and in the positive direction, so $\gamma_0 \cdot \mu = 1$. $\qquad \square$

It should be remarked that the definition of α does not depend on the choice of $M \in \Lambda(E)$, because the connected group $\mathrm{Sp}(E, \mathbb{R})$ acts transitively on $\Lambda(E)$ and the intersection number is homotopy-invariant.

That $\mathrm{Sp}(E)$ acts transitively on $\Lambda(E)$ follows from the fact that we can for every $L, M, \tilde{L}, \tilde{M} \in \Lambda(E)$ such that $L \cap M = 0$, $\tilde{L} \cap \tilde{M} = 0$, find a symplectic mapping A sending L into $\tilde{L}$, M into $\tilde{M}$. We can prescribe A arbitrarily (but bijective) on L, and then A is uniquely determined on M. (Apply Theorem 3.4.2 and then A is the canonical transformation induced by the mapping: $L \to \tilde{L}$.)

To see that $\text{Sp}(E)$ is connected, we remark that $\text{Sp}(E)$ is a fiber bundle over the connected space $\Lambda(E)$, with fiber over $L \in \Lambda(E)$ equal to the group H of symplectic transformations leaving L invariant. H is isomorphic to $\text{GL}(n, R) \times R^{(1/2)n(n+1)}$, in view of Theorem 3.4.3 and therefore has two components, the component of the identity and the component of the matrix that maps e_1 to $-e_1$, f_1 to $-f_1$, and leaves the other basis vectors fixed. Now the group of symplectic transformations sending the linear span P of e_1 and f_1 into itself and leaving the other basis vectors fixed is isomorphic to the group of symplectic transformations of the symplectic plane (P, σ). But this group is connected, so the two components of H can be connected in $\text{Sp}(E)$.

Note that a closed curve γ is not homotopic to zero if and only if it has nonzero intersection number with $\Lambda(E) \setminus \Lambda^0(E, M)$ for some $M \in \Lambda(E)$ and then any curve $\tilde{\gamma}$ homotopic to γ has nonzero intersection with $\Lambda(E) \setminus \Lambda^0(E, \tilde{M})$, for every $\tilde{M} \in \Lambda(E)$.

Whereas $\Lambda^0(E, M)$ is diffeomorphic to a vector space, the intersection $\Lambda^0(E, M_1) \cap \Lambda^0(E, M_2)$ is in general no longer connected. The following quantity measures whether L_1, L_2 are in the same component of $\Lambda^0(E, M_1) \cap \Lambda^0(E, M_2)$ or not.

Definition 3.4.2. Let L_1, L_2, M_1, M_2 be four Lagrangian subspaces of (E, σ) such that L_j is transversal to M_k for all $j, k = 1, 2$. Then the *Hörmander index* is defined by

$$(3.4.15) \qquad\qquad s(M_1, M_2; L_1, L_2) = \langle \gamma, \alpha \rangle,$$

where γ is a closed curve in $\Lambda(E)$ consisting of an arc of Lagrangian subspaces from L_1 to L_2 transversal to M_1, followed by an arc of Lagrangian subspaces from L_2 to L_1 transversal to M_2. Here α is the Maslov–Arnol'd index.

Proposition 3.4.10. $s(M_1, M_2; L_1, L_2) = 0$ *if and only if L_1 and L_2 are in the same connected component of* $\Lambda^0(E, M_1) \cap \Lambda^0(E, M_2)$.

Proof. A curve γ in $\Lambda^0(E, M_1) \cup \Lambda^0(E, M_2)$ is modulo $\pi_1(\Lambda^0(E, M_1) \cup \Lambda^0(E, M_2))$ homotopic to zero in $\Lambda^0(E, M_1)\Lambda^0(E, M_2)$ if and only if $\langle \gamma, \alpha \rangle = 0$. The proof of Theorem 3.4.9 applies with no change here because $\Lambda^1(E, M_1) \cap \Lambda^0(E, M_2)$ is connected: $L \mapsto L \cap M_1$ is a fibration of this space over the lines ℓ in M_1 transversal to $M_1 \cap M_2$, with fiber over ℓ isomorphic to $\Lambda^0(\ell^\sigma/\ell, (M_2 \cap \ell^\sigma)/\ell)$. Both base space and fiber are connected. So if $\langle \gamma, \alpha \rangle = 0$ then we can replace γ by another curve as in Definition 3.4.2 that is homotopic to zero in $\Lambda^0(E, M_1) \cup \Lambda^0(E, M_2)$.

So we must show that L_1, L_2 are in the same connected component of $\Lambda^0(E, M_1) \cap \Lambda^0(E, M_2)$ if and only if γ is homotopic to zero in $\Lambda^0(E, M_1) \cup \Lambda^0(E, M_2)$. The "if" part follows because $\Lambda^0(E, M_1)$ and $\Lambda^0(E, M_2)$ are open in $\Lambda^0(E, M_1) \cup \Lambda^0(E, M_2)$, the converse is true because $\Lambda^0(E, M_1)$ and $\Lambda^0(E, M_2)$ are simply connected. $\qquad\square$

Clearly s is integer-valued and continuous (= locally constant) in all variables. Moreover:

$$(3.4.16) \quad s(M_1, M_2; L_1, L_2) = -s(M_1, M_2; L_2, L_1) = s(M_2, M_1; L_2, L_1)$$

(antisymmetry in L_1, L_2 and in M_1, M_2) and

$$(3.4.17) \quad s(M_1, M_2; L_1, L_2) + s(M_1, M_2; L_2, L_3) = s(M_1, M_2; L_1, L_3)$$

(cocycle condition in the L's).

We are now going to compute s.

Definition 3.4.3. If M_1, M_2 are transversal Lagrangian subspaces, L transversal to M_2, then $\text{sgn}\,(M_1, M_2; L)$ is the signature of the symmetric bilinear form $Q(L): (x, y) \to \sigma(Ax, y)$ on M_1, if $L = \{x + Ax; x \in M_1\}$, $A: M_1 \to M_2$.

Lemma 3.4.10. *If M_1, M_2 are transversal (and the L_j transversal to the M_k), then*

$$(3.4.18) \quad s(M_1, M_2; L_1, L_2) = \tfrac{1}{2}[\text{sgn}\,(M_1, M_2; L_1) - \text{sgn}\,(M_1, M_2; L_2)].$$

Proof. $s(M_1, M_2; L_1, L_2)$ is clearly equal to the intersection number of a differentiable curve $\gamma(t)$ from L_2 to L_1 in $\Lambda^0(E, M_2)$ with $\Lambda(E)\backslash\Lambda^0(E, M_1)$, intersecting the latter only in its regular part $\Lambda^1(E, M_1)$ and transversally. Now $\text{sgn}\,(M_1, M_2; \gamma(t))$ only changes when $\gamma(t)$ crosses $\Lambda^1(E, M_1)$. We claim that it changes by $+2$, respectively, -2, if the crossing is positive, respectively, negative.

Suppose the crossing at $t = t_0$ is positive. Choose coordinates in M_1 such that $\gamma(t_0) \cap M_1$ is the x_1-axis and

$$Q(\gamma(t_0)) = \begin{pmatrix} 0 & 0 \\ 0 & Q_0 \end{pmatrix}$$

for some symmetric, nonsingular $(n-1) \times (n-1)$ matrix Q_0. Now

$$(3.4.19) \qquad \operatorname{sgn} \begin{pmatrix} B & C \\ C' & D \end{pmatrix} = \operatorname{sgn} D + \operatorname{sgn}[B - CD^{-1}C']$$

if D is nonsingular, in view of the formula

$$\begin{pmatrix} I & -CD^{-1} \\ 0 & I \end{pmatrix} \begin{pmatrix} B & C \\ C' & D \end{pmatrix} \begin{pmatrix} I & 0 \\ -(D')^{-1}C' & I \end{pmatrix} = \begin{pmatrix} B - CD^{-1}C' & 0 \\ 0 & D \end{pmatrix}.$$

Here $CD^{-1}C'$ vanishes of second order as $C = 0$, so

$$\operatorname{sgn}[B(t) - C(t) \cdot D(t)^{-1}C(t)']$$

jumps by $+2$ if t passes t_0, $B(t_0) = C(t_0) = 0$, $D(t_0) = Q_0$, $B'(t_0)$ positive. Note that $\operatorname{sgn} D(t)$ remains constant.

Finally we remark that

$$\gamma(t) = \{x + A(t)x; \ x \in M_1\}, \quad A(t): M_1 \to M_2$$
$$= \{y + \tilde{A}(t)y; \ y \in \gamma(t_0)\}, \quad \tilde{A}(t): \gamma(t_0) \to M_2.$$

So $A(t) = A(t_0) + \tilde{A}(t)(I + A(t_0))$ on M_1, hence

$$(3.4.20) \qquad \frac{d}{dt}\sigma(A(t)x, y)_{t=t_0} = \frac{d}{dt}\sigma(\tilde{A}(t)x, y)_{t=t_0}$$

if $x, y \in \gamma(t_0) \cap M_1$. So $B'(t_0)$ is positive if and only if $\gamma(t)$ crosses $\Lambda^1(E, M_1)$ in the positive direction.   □

Lemma 3.4.11. $\operatorname{sgn}(M_1, M_2; L) = -\operatorname{sgn}(M_1, L; M_2)$ *if* M_1, M_2, L *are mutually transversal.*

Proof. We can write

$$L = \{x + Ax; x \in M_1\}, \quad A: M_1 \to M_2$$
$$M_2 = \{y + By; y \in M_1\}, \quad B: M_1 \to L.$$

If $x \in M_1$, then $(x + Ax) + Bx = (x + Bx) + Ax$ belongs to both L and M_2, hence $x + Ax + Bx = 0$. So

$$\sigma(Ax, y) = -\sigma(Bx, y), \qquad x, y \in M_1.$$ □

Theorem 3.4.12. *If* $M_1, M_2, L_1, L_2 \in \Lambda(E)$, L_j *transversal to* M_k *for* $j, k = 1, 2$, *then*

$$(3.4.21) \quad s(M_1, M_2; L_1, L_2) = \tfrac{1}{2}[\operatorname{sgn}(M_1, L_2; M_2) - \operatorname{sgn}(M_1, L_1; M_2)].$$

Proof. Note that the right-hand side is well-defined and that (3.4.21) holds if M_1, M_2 are transversal in view of Lemmas 3.4.10 and 3.4.11. Now assume dim $M_1 \cap M_2 = k$; let $\gamma(t)$ be a curve in $\Lambda(E)$ such that $\gamma(0) = M_2$ and $\gamma'(0)$, regarded as a quadratic form on M_2, is positive definite on $M_1 \cap M_2$. Then, using formulas analogous to (3.4.19), (3.4.20) we obtain that $\gamma(t)$ is transversal to M_1,

$$\mathrm{sgn}\,(M_1, L_2; \gamma(t)) = \mathrm{sgn}\,(M_1, L_2; M_2) + k$$
$$\mathrm{sgn}\,(M_1, L_1; \gamma(t)) = \mathrm{sgn}\,(M_1, L_1; M_2) + k$$

if $t > 0$ is small. This proves (3.4.21) because it holds M_2 is replaced by $\gamma(t)$, $t > 0$, small, and the left-hand side is locally constant in M_2. $\qquad\square$

Remark. $\mathrm{sgn}\,(M_1, L; M_2)$ is not continuous in M_1, M_2, L, it is only continuous if we restrict to triples such that L is transversal to M_1 and M_2, and dim $M_1 \cap M_2$ is constant. Because s is expressed in terms of signatures of quadratic forms it is sometimes also referred to as a "Morse index." For more information regarding $\Lambda(E)$, not found in Arnol'd [5] or Hörmander [40], 3.3, see Fuks [30]. (This article discusses rather special questions.)

3.5. Symplectic differential geometry

Suppose (M, σ), (N, τ) are symplectic manifolds. A *symplectic mapping* from (M, σ) to (N, τ) is a differentiable mapping Φ such that $\Phi^* \tau = \sigma$, that is, $D\Phi_m$ is a symplectic linear mapping from $(T_m(M), \sigma_m)$ to $(T_{\Phi(m)}(N), \tau_{\Phi(m)})$ for all $m \in M$. This implies that Φ is an immersion, so dim $M \leq$ dim N and Φ is a local diffeomorphism if dim $M =$ dim N. If $N = T^*(X)$, $\tau =$ canonical 2-form, then Φ is also called a *canonical mapping*, and a *canonical coordinatization* if $X = \mathbb{R}^n$. Example: if κ is a local diffeomorphism: $X \to Y$ then the *induced mapping* $\tilde{\kappa}: T^*(X) \to T^*(Y)$, defined by $\tilde{\kappa}(x, \xi) = (\kappa(x), ({}^t D\kappa_x)^{-1} \cdot (\xi))$, is a canonical transformation. See the remark after Theorem 3.4.2.

Definition 3.5.1. Let v be a C^∞ vector field on a symplectic manifold (M, σ), Φ^t its flow on M. Then v is called a *symplectic vector field* or an *infinitesimal symplectic transformation*, if the Φ^t are symplectic transformations: $(U_t, \sigma) \to (M, \sigma)$.

Theorem 3.5.1. *The following assertions are equivalent.*

  (i) v *is symplectic*
  (ii) $\mathcal{L}_v\sigma = 0$
  (iii) $d(\sigma \mid v) = 0$, *that is, $\sigma \mid v = df$ (locally) for some C^∞ function f.*

Proof. (i) means that $(\Phi^t)^*\sigma = \sigma$ for all t, which is evidently equivalent to (ii), in view of the definition (3.2.15) of the Lie-derivative. The equivalence with (iii) follows from Theorem 3.2.5, remarking that $d\sigma = 0$, and the Poincaré lemma. $\qquad\square$

Definition 3.5.2. For any $f \in C^\infty(M)$ the *Hamilton field* H_f is the unique C^∞ vector field v on M such that $df = -(\sigma \mid v)$.

Here it is used that σ_m is an isomorphism: $T_m(M) \to T_m(M)^*$ for each $m \in M$, in view of the nondegeneracy of σ. Note that each Hamilton field is symplectic and that conversely each symplectic vector field v is locally the Hamilton field of some function f, which then is called a *Hamilton function* for v. Note that the Hamilton function is unique modulo a constant, and that a global Hamilton function always can be found if and only if $H^1(M,\mathbb{R}) = 0$ (see Theorem 3.2.7), which in turn is the case if M is simply connected.

Theorem 3.5.2 (Jacobi). *Suppose (M,σ) and (N,τ) are symplectic manifolds of the same dimension and let Φ be a symplectic mapping: $(M,\sigma) \to (N,\tau)$, then*

(3.5.1) $\Phi^* H_f = H_{\Phi \cdot f}$ *for every $f \in C^\infty(N)$.*

Conversely, if (3.5.1) holds for $f = f_j$ where the $d(\Phi^* f_j)_m$ span $T_m(M)^*$ for all $m \in M$, then Φ is symplectic.

Proof. In view of (3.2.19), (3.2.14) we have $\Phi^*\tau \mid \Phi^* H_f = \Phi^*(\tau \mid H_f) = -\Phi^* df = -d\Phi^* f$. So $\Phi^*\tau = \sigma$ implies (3.5.1), while conversely (3.5.1) implies $\Phi^*\tau = \sigma$ when taken inner product with $\Phi^* H_f$. Note that if the $d(\Phi^* f_j)_m$ span $T_m(M)^*$, then $D\Phi_m$ is injective. Also the $(df_j)_n$ must span $T_n(N)^*$, which implies that the $H_{f_j}(n)$ span $T_n(N)$. So the $\Phi^* H_{f_j}(m)$ span $T_m(M)$, and $(\Phi^*\tau - \sigma)_m = 0$ on them implies $\Phi^*\tau = \sigma$. $\qquad\square$

Definition 3.5.3. If $f,g \in C^\infty(M)$ then the *Poisson brackets* $\{f,g\} \in C^\infty(M)$ are defined by

(3.5.2) $\{f,g\} = H_f g = \sigma(H_f, H_g)$.

Corollary 3.5.3. *The functions $x_1, \ldots, x_n$, $\xi_1, \ldots, \xi_n$ form a canonical coordinatization of (M, σ), dim $M = 2n$, if and only if for all $i, j = 1, \ldots, n$:*

$$(3.5.3) \qquad \{x_i x_j\} = 0, \quad \{\xi_i, \xi_j\} = 0, \quad \{\xi_i, x_j\} = \delta_{ij}.$$

Proof. Follows immediately from Theorem 3.5.2 and the remark that (3.5.3) hold for the coordinate functions $x_1, \ldots, x_n, \xi_1, \ldots, \xi_n$ in $\mathbb{R}^n \times (\mathbb{R}^n)^*$, where $\mathbb{R}^n \times (\mathbb{R}^n)^*$ is provided with the canonical 2-form. Also note that (3.5.3) implies that the $dx_i, \ldots, dx_n, d\xi_1, \ldots, d\xi_n$ are linearly independent of each point. For the Hamilton fields we obtain the formula

$$(3.5.4) \qquad H_f = \sum_{j=1}^{n} \left(\frac{\partial f}{\partial \xi_j} \frac{\partial}{\partial x_j} - \frac{\partial f}{\partial x_j} \frac{\partial}{\partial \xi_j} \right)$$

for a differentiable function f on $\mathbb{R}^n \times (\mathbb{R}^n)^*$. (Compare this also with (2.5.7).) $\qquad\square$

Theorem 3.5.4. *For $f, g \in C^\infty(M)$ we have*

$$(3.5.5) \qquad [H_f, H_g] = H_{\{f,g\}}.$$

Proof. Let Φ^t be the H_f-flow. Then, using (3.1.8) and the fact that the Φ^t are symplectic, we have:

$$[H_f, H_g] = \frac{d}{dt}(\Phi^t)^* H_g \big|_{t=0} = \frac{d}{dt} H_{(\Phi^t)^* g} \big|_{t=0} = H_{H_f g}. \qquad\square$$

Corollary 3.5.5. $C^\infty(M)$ *is a Lie-algebra with the Poisson brackets product.*

Proof. The antisymmetry follows from (3.5.2) and the antisymmetry of σ. For the Jacobi-identity we remark that

$$\{f, \{g, h\}\} + \{g, \{h, f\}\} + \{h, \{f, g\}\} = (H_f \circ H_g - H_g \circ H_f) h - H_{\{f,g\}} h = 0$$

in view of (3.1.24) and (3.5.5). $\qquad\square$

The following theorem expresses that all symplectic manifolds of the same dimension are locally symplectically isomorphic. (This is very different from the situation with Riemannian manifolds!)

Theorem 3.5.6 (Darboux). *Suppose (M, σ) is a symplectic manifold. Then* dim M *is even, say* $2n$, *and for each* $m_0 \in M$ *there is a canonical coordinatization of a neighborhood* U *of* m_0.

The main step in the proof is the following.

Theorem 3.5.7 (Jacobi). *Suppose the C^∞ functions $\xi_1, \ldots, \xi_k$ on a neighborhood U of m_0 are in involution, that is*

(3.5.6)

$\quad$ (i) $d\xi_1, \ldots, d\xi_k$ *are linearly independent at each* $m \in U$, *and*

$\quad$ (ii) $\{\xi_i, \xi_j\} = 0$ *for all* $i, j = 1, \ldots, k$.

Then $k \le n$ and there exists a neighborhood $\tilde{U}$ of m_0 and $\xi_{k+1}, \ldots, \xi_n \in C^\infty(\tilde{U})$ such that $\xi_1, \ldots, \xi_n$ are in involution on $\tilde{U}$.

Proof. The linear independence of the $d\xi_1, \ldots, d\xi_k$ is equivalent to the linear independence of the $H_{\xi_1}, \ldots, H_{\xi_k}$. The condition (ii) means in view of (3.5.2) that the span S_m of the $H_{\xi_i}(m)$, $1 \le i \le k$, is an isotropic subspace of $(T_m(M), \sigma_m)$, hence $k \le n$ (see Proposition 3.4.1).

Now let $k < n$. The problem is whether we can find a function $\xi = \xi_{k+1}$ such that the $\xi_1, \ldots, \xi_k, \xi_{k+1}$ are in involution. We have automatically $\{\xi, \xi\} = 0$ so we only need to solve

(3.5.7) $\qquad\qquad \{\xi_i, \xi\} = H_{\xi_i}\xi = 0, \qquad i = 1, \ldots, k,$

with a solution ξ such that $d\xi$ is linearly independent of the $d\xi_i$.

Now (ii) implies, because of (3.5.5), that the Hamilton fields H_{ξ_i} commute, so in view of the Frobenius theorem (Theorem 3.1.1) we can choose local coordinates on which $H_{\xi_i} = \partial/\partial x_i$ (we do not need these local coordinates to be canonical). We therefore can solve (3.5.7) and even prescribe ξ on any manifold T that is transversal to the span S_m of the H_{ξ_i}, that is, $T_{m_0}(M) = T_{m_0}(T) \oplus S_{m_0}$.

Now choose $u \in (S_{m_0})^\sigma$, $u \notin S_{m_0}$ (see the proof of Proposition 3.4.1), choose T such that $u \in T_{m_0}(T)$ and finally ξ such that $d\xi_{m_0}(u) \ne 0$. This means that $H_\xi(m_0)$ is not σ-orthogonal to a vector u to which each $H_{\xi_i}(m_0)$ is σ-orthogonal, so $d\xi_{m_0}$ is linearly independent from the $(d\xi_i)_{m_0}$. $\qquad\square$

Proof of the Darboux theorem.* In view of Theorem 3.5.7 we can construct $\xi_1, \ldots, \xi_n$, which are in involution on a neighborhood of m_0. Suppose we have in addition a set of functions $x_1, \ldots, x_k$ such that

$$\{x_i, x_j\} = 0 \qquad \text{for all } i, j = 1, \ldots, k$$
$$\{\xi_i, x_j\} = \delta_{ij} \qquad \text{for all } i = 1, \ldots, n, \ j = 1, \ldots, k.$$

These equations imply that the $H_{\xi_1}, \ldots, H_{\xi_n}, H_{x_1}, \ldots, H_{x_k}$ are linearly independent, and commutating so we can find $x = x_{k+1}$ such that these equations hold with k replaced by $k+1$. So we finally find $\xi_1, \ldots, \xi_n$, $x_1, \ldots, x_n$ satisfying (3.5.3). $\qquad\square$

Knowing that any symplectic manifold admits local canonical coordinates, the volume introduced in the following theorem gets more meaning.

Theorem 3.5.8. *Let (M, σ) be a symplectic manifold*, $\dim M = 2n$. *Define*

$$(3.5.8) \qquad \omega = \frac{1}{n!}(-1)^{\binom{n}{2}} \sigma \wedge \sigma \cdots \wedge \sigma \quad (n \text{ times}).$$

Then

$$(3.5.9) \qquad \omega = \kappa^*(d\xi_1 \wedge \cdots \wedge d\xi_n \wedge dx_1 \wedge \cdots \wedge dx_n)$$

for each local canonical coordinatization κ. ω is called the canonical volume of (M, σ).

Proof. In $\mathbb{R}^n$ we have

$$\left(\sum_{i=1}^{n} d\xi_i \wedge dx_i\right) \wedge \cdots \wedge \left(\sum_{i=1}^{n} d\xi_i \wedge dx_i\right) \quad (n \text{ times})$$
$$= n!(-1)^{\binom{n}{2}} d\xi_1 \wedge \cdots \wedge d\xi_n \wedge dx_1 \wedge \cdots \wedge dx_n.$$

Theorem 3.5.9 (Liouville). *Let Φ^t be the flow of a Hamilton field on a symplectic manifold (M, σ). Then $(\Phi^t)^*\omega = \omega$ for all t, if ω denotes the canonical volume of (M, σ).*

Proof. Use $(\Phi^t)^*\sigma = \sigma$ and (3.2.13). $\qquad\square$

---

* A simpler proof is given by Weinstein [81], but we need Theorem 3.5.7 in the applications anyhow.

The theorems of Darboux and Jacobi are in general only locally valid. The following theorem shows how restrictive the existence of a global set of functions $f_1, \ldots, f_n$ in involution really is.

Theorem 3.5.10 (Arnol'd [1]). *Suppose (M, σ) is a symplectic manifold of dimension $2n$, let $f_1, \ldots, f_n$ be in involution on M, and finally assume that the Hamilton fields $H_{f_1}, \ldots, H_{f_n}$ are complete, that is, their solution curves are defined for all $t \in \mathbb{R}$.*

Then $f: m \to (f_1(m), \ldots, f_n(m))$ defines a submersion from M onto an open subset of $\mathbb{R}^n$. Each connected component of each fiber is diffeomorphic with a cylinder $T^k \times \mathbb{R}^{n-k}$, here $T^k = \mathbb{R}^k / \mathbb{Z}^k$ is the k-dimensional torus, and

$$\Phi: (t_1, \ldots, t_n) \to \Phi_n^{t_n} \circ \cdots \circ \Phi_1^{t_1}$$

is a transitive group action of $\mathbb{R}^n$ on it. Here Φ_i^t denotes the H_{f_i}-flow.

Proof. The linear independence of the df_i implies that f defines a submersion. In particular the fibers $V_c = \{m \in M; \ f(m) = c\}$ are n-dimensional smooth submanifolds of M. The V_c are invariant under the H_{f_i}-flow because $H_{f_i} f_j = \{f_i, f_j\} = 0$.

Because the H_{f_i} commute, also the Φ^{t_i} commute in view of Frobenius' theorem. So Φ is a group action of $\mathbb{R}^n$ on each connected component Γ of V_c. The image of a point under the action Φ is open because the H_{f_i} span the tangent space of Γ (use the implicit function theorem). Its complement, being Φ-invariant, is also open, and we conclude that Φ is transitive on Γ.

For a transitive action of any Lie group G on a topological space Γ it is known that Γ is diffeomorphic to G/S_m, where S_m is the *stationary group* or *isotropy group*

$$S_m = \{g \in G; \ g(m) = m\}.$$

In our case, using again the implicit function theorem, we find that S_m is discrete, which implies that Γ has the form $T^k \times \mathbb{R}^{n-k}$.

Remark. The submersion f need not be a fibration. Example: Take $n = 1$ and $f(x, \xi) = \frac{1}{2}\xi^2 - \cos x$, the total energy function of the mathematical pendulum, defined on

$$M = ((\mathbb{R}/2\pi\mathbb{Z}) \times \mathbb{R}) \setminus \{(0, 0), (\pi, 0)\},$$

the complement in the phase space of the set of equilibrium points. On the open H_f-invariant subsets

$$M_{\neq 1} = \{(x,\xi) \in M; \ f(x,\xi) \neq 1\},$$

f is a fibration with fibers diffeomorphic to circles, whereas the fiber $f = 1$ is diffeomorphic to two copies of $\mathbb{R}$.

In general one can prove, by means of the implicit function theorem, that if a connected component Γ of V_{c_0} is compact, hence diffeomorphic to T^n, then the $S_{m(c)} \subset \mathbb{R}^n$ depend smoothly on c, where we can arrange $m(c) \in V_c$ to depend smoothly on c, for c near c_0 and $m(c_0) \in \Gamma$. In this case f is a smooth fibration with toral fibers, on an H_{f_i}-invariant open neighborhood of Γ in M. The problem with noncompact Γ is that elements of S_m can "enter from infinity".

In the classical literature (see Whittaker [82], p. 323) the flow of a Hamilton field H_f is called *completely integrable* if there exist global functions $f_2, \ldots, f_n$ such that $f, f_2, \ldots, f_n$ are in involution. This terminology is consistent with the definition of an *integral* for the H_f-flow as any H_f-invariant function g such that its differential is linearly independent of df (and df never vanishes), only if $n = 2$. For $n > 2$ the existence of linearly independent integrals $f_2, \ldots, f_n$ is not sufficient for complete integrability since we also need the relations $\{f_i, f_j\} = 0$ for $i \neq j$, $i,j = 2, \ldots, n$.

A curve $t \to \gamma(t)$ on a manifold M is called *quasiperiodic* if it is contained in a submanifold V of M for which there exists a diffeomorphism $\Phi\colon V \to T^d$, and a constant vector field w on T^d, such that $\Phi \circ \gamma$ has constant velocity in T^d. Note that if $k = n$ in Theorem 3.5.10 then each H_{f_i} has only quasiperiodic solution curves. It is a deep result of Kolmogorov–Arnol'd [6] and Moser [66], that small Hamiltonian perturbations of an integrable vector field H_f (with compact fibers of $(f_1, f_2, \ldots, f_n)$ and with "generically moving" S_m) still have many invariant tori on which the motion is quasiperiodic.

3.6. Lagrangian manifolds

Definition 3.6.1. Let (M, σ) be a symplectic manifold, $\dim M = 2n$. A submanifold V of M is called *isotropic, Lagrangian,* and *involutive* in (M, σ) if $T_m(V) \subset T_m(V)^\sigma$, $T_m(V) = T_m(V)^\sigma$, and $T_m(V)^\sigma \subset T_m(V)$ for all $m \in V$, respectively.

Note that if V is isotropic, Lagrangian, and involutive then $\dim V \leq n$, $\dim V = n$, and $\dim V \geq n$, respectively. Conversely if V is isotropic or involutive and dim $V = n$ then V is Lagrangian (see Proposition 3.4.1). The following lemma explains the relation with "involutive" in Theorem 3.5.7.

Lemma 3.6.1. *Let V locally be defined by $f_1 = \cdots = f_k = 0$, $f_j \in C^\infty(M)$, df_j linearly independent at V. Then V is involutive if and only if*

$$(3.6.1) \quad \{f_i, f_j\} = 0 \quad on \quad f_1 = \cdots = f_k = 0 \quad for \ all \ \ i, j = 1, \ldots, k.$$

Proof.

$$T_m(V) = \bigcap_{i=1}^{k} \ker (df_i)_m = \bigcap_{i=1}^{k} [H_{f_i}(m)]^\sigma = \left(\sum_{i=1}^{k} \mathbb{R} \cdot H_{f_i}(m)\right)^\sigma,$$

so $T_m(V)^\sigma = \sum\limits_{i=1}^{k} \mathbb{R} \cdot H_{f_i}(m)$. Therefore $T_m(V)^\sigma \subset T_m(V)$ just means that the H_{f_i} are tangent to V, which is exactly (3.6.1). □

Theorem 3.6.2. *Let V be a C^∞ submanifold of codimension k in M. Through each $m \in V$ there passes a Lagrangian manifold $L \subset V$ if and only if V is involutive.*

If V is involutive then the $T_m(V)^\sigma$, $m \in V$ form an integrable tangent system in V, the integral manifolds of which are called the characteristic strips of V. A submanifold M_0 of codimension k in V is symplectic if and only if it is transversal to the characteristic strips. If this is the case then for any Lagrangian manifold L_0 in (M_0, σ) there is locally exactly one Lagrangian manifold L in (M, σ) such that $L_0 \subset L \subset V$, L is equal to the union of the characteristic strips passing through points of L_0.

Proof. If through $m \in V$ there passes a Lagrangian manifold $L \subset V$ then

$$(3.6.2) \qquad T_m(v) \supset T_m(L) = T_m(L)^\sigma \supset T_m(V)^\sigma.$$

This shows that V is involutive, if this happens for each $m \in V$.

Now suppose V is involutive. Representing it locally by $f_1 = \cdots = f_k = 0$, with the df_i linearly independent, we see from 3.6.1 that $d\{f_i, f_j\}$ is a linear combination of the df_ℓ (at V), hence $[H_{f_i}, H_{f_j}] = H_{\{f_i, f_j\}}$ is a linear combination of the H_{f_ℓ}. So the $T_m(V)^\sigma$ form an integrable tangent system according to Frobenius' theorem (Th. 3.1.1).

Now let M_0 be a submanifold of codimension k in V. Then (M_0, σ) is symplectic if and only if $T_m(M_0) \cap T_m(M_0)^\sigma = (0)$. Now $T_m(M_0) \subset T_m(V)$ so $T_m(V)^\sigma \subset T_m(M_0)^\sigma$ so necessarily $T_m(M_0) \cap T_m(V)^\sigma = (0)$. Conversely, if $T_m(M_0) \cap T_m(V)^\sigma = (0)$ then $T_m(M_0)^\sigma + T_m(V) = T_m(M)$ so $T_m(M_0)^\sigma \cap T_m(V)$ is k-dimensional, hence $T_m(V)^\sigma = T_m(M_0)^\sigma \cap T_m(V)$ and $T_m(M_0) \cap T_m(M_0)^\sigma = (0)$.

Now let M_0 be of codimension k in V and transversal to the characteristic strips, and let L_0 be a Lagrangian submanifold of (M_0, σ). In view of (3.6.2) the characteristic strips are tangent to any Lagrangian submanifold $\tilde{L}$ of V so if in addition $L_0 \subset \tilde{L}$ then $\tilde{L}$ contains the characteristic strips passing through points of L_0 (locally). Now define L locally as the union of the local characteristic strips through points of L_0, more precisely:

$$(3.6.3) \qquad L = \left\{ \Phi_k^{t_k} \circ \cdots \circ \Phi_1^{t_1}(m); \ m \in U \cap L_0, \ (t_1, \ldots, t_k) \in T \right\},$$

where U is a neighborhood of $m_0 \in L_0$ on which V is defined by $f_1 = \cdots = f_k = 0$ with linearly independent df_i, Φ_i^t is the H_{f_i}-flow and T is a neighborhood of the origin in $\mathbb{R}^k$.

Then the differential at $(0, m_0)$ of the mapping: $(t_1, \ldots, t_k, m) \mapsto \Phi_k^{t_k} \circ \cdots \circ \Phi_1^{t_1}(m)$: $T \times (U \cap L_0)$ has rank n because of the transversality of $T_{m_0}(L_0)$ and $T_{m_0}(V)^\sigma$, so we conclude that for T, U sufficiently small, the mapping is an embedding and the image L is an n-dimensional smooth submanifold of V, $U \cap L_0 \subset L$. Because $\tilde{L} \supset L$ (locally) and $\dim \tilde{L} = \dim L$, we conclude the local uniqueness in the theorem.

To prove the existence part we only have to show that L is actually Lagrangian. Define

$$(3.6.4) \qquad L_1 = \left\{ \Phi_1^{t_1}(m); \ m \in U \cap L_0, \ t_1 \in T_1 \right\},$$

where T_1 is a sufficiently small interval in $\mathbb{R}$ around 0. Then, for $m = \Phi_1^0(m) \in U \cap L_0$, we have

$$T_m(L_1) = T_m(L_0) + \mathbb{R} \cdot H_{f_1}(m),$$

so

$$T_m(L_1)^\sigma = T_m(L_0)^\sigma \cap \ker df_{1,m} \supset T_m(L_0) + \mathbb{R} \cdot H_{f_1}(m)$$

because $T_m(L_0) \subset T_m(V)$ implies $T_m(V)^\sigma \subset T_m(L_0)^\sigma$ we conclude that $T_m(L_1)$ is isotropic if $m \in U \cap L_0$. However, $T_{\Phi_1^t(m)}(L_1) = D\Phi_{1,m}^t(T_m(L_1))$ is then also isotropic because $D\Phi_{1,m}^t$ is a symplectic linear mapping: $T_m(M) \to T_{\Phi_1^t(m)}(M)$. So we conclude that L_1 is isotropic. Repeating the arguments we find that

$$(3.6.5) \qquad L_\ell = \left\{ \Phi_1^{t_\ell} \cdots \Phi_1^{t_1}(m), \ m \in U \cap L_0, \ (t_1, \ldots, t) \in T_\ell \right\}$$

is an isotropic submanifold of dimension $n-k+\ell$ if U and T_ℓ are sufficiently small neighborhoods of m_0 and the the origin in R^ℓ, respectively. The building-up stops at $\ell = k$ where we find a Lagrangian submanifold $L = L_k$ through m_0', $U \cap L_0 \subset L \subset V$. Piecing together these local solutions along L_0 we get a Lagrangian manifold L such that $L_0 \subset L \subset V$, in view of the local uniqueness this involves no problems. If one admits immersed submanifolds in V, then one can also conclude that there exists a unique maximal immersed Lagrangian manifold L in V containing L_0.   □

Application of Theorem 3.6.2 to the case that $M = T^*(X)$, V given by equation $f_1 = \cdots = f_k = 0$, we obtain the following theorem on (nonlinear!) first-order differential equations of the form

$$(3.6.6) \qquad f_1(x, d\phi_x) = \cdots = f_k(x, d\phi_x) = 0.$$

The result is usually referred to as *Hamilton–Jacobi theory*, but we follow Lie's presentation.

Theorem 3.6.3. *Let $f_1, \ldots, f_k$ be real C^r functions on an open subset Ω of $T^*(X)$, $2 \le r \le \infty$, such that $df_1, \ldots, df_k$ are linearly independent at*

$$V = \{(x, \xi) \in \Omega; \ f_1(x, \xi) = \cdots = f_k(x, \xi) = 0\}.$$

For every $(x, \xi) \in V$ there exists a C^2 solution ϕ of (3.6.6) on a neighborhood of x and such that $d\phi_x = \xi$, only if the $f_1, \ldots, f_k$ are in involution and

$$(3.6.7) \qquad d_\xi f_1, \ldots, d_\xi f_k \quad \text{are linearly independent at } V.$$

If conversely these conditions are satisfied then for any $(x_0, \xi_0) \in V$, any $(n-k)$-dimensional submanifold Q of X through x_0 that is transversal to the linear span of the $d_\xi f_i(x_0, \xi_0)$, $i = 1, \ldots, k$, and any C^r function ψ on Q such that $d\psi_{x_0} = \xi_0\big|_{T_{x_0}(Q)}$, there is a locally unique C^r solution ϕ such that

$$(3.6.8) \qquad\qquad \phi = \psi \quad \text{on } Q.$$

Moreover, the graph of $d\phi$ in $T^*(X)$ is a union of bicharacteristic strips, that is, it is invariant under the H_{f_i}-flows, $i = 1, \ldots, k$.

Proof. If ϕ is a C^2 solution then the graph L of $d\phi$ in $T^*(X)$ is a Lagrangian C^1 manifold contained in V, such that $D\pi_{(x,\xi)}$: $T_{(x,\xi)}(L) \to T_x(X)$ is injective. See Proposition 3.3.2. Because

$$T_{(x,\xi)}(L) \supset T_{(x,\xi)}(V)^\sigma = \sum_{i=1}^k \mathbb{R} \cdot H_{f_i}(x,\xi)$$

it follows that the $D\pi_{(x,\xi)}H_{f_i}(x,\xi) = d_\xi f_i(x,\xi)$ are linearly independent for $i = 1,\ldots,k$. It follows also that V must be involutive.

Now assume that V is involutive and (3.6.7) holds, and let Q, ψ be as above. It follows then from the implicit function theorem that the equations

(3.6.9) $$\xi\big|_{T_x(Q)} = d\psi_x, \qquad f_1(x,\xi) = \cdots = f_k(x,\xi) = 0$$

for every $x \in Q$ have exactly one solution $\xi_x \in T_x(X)^*$ locally (close to ξ_0), and that ξ_x depends C^{r-1} on x.

Its graph $L_0 = \{(x,\xi_x) \in T^*(X); x \in Q\}$ is an $(n-k)$-dimensional isotropic submanifold of V (because $\{(x,\xi) \in T^*(X);\ x \in Q,\ \xi\big|_{T_x(Q)} = d\psi(x)\}$ is a Lagrange manifold) that is transversal to the characteristic strips of V. Application of Theorem 3.6.2 leads to a locally unique Lagrangian C^{r-1} submanifold L in V, containing L_0. $D\pi_{(x_0,\xi_0)}$ maps $T_{(x_0,\xi_0)}(L_0)$ onto $T_{x_0}(Q)$ and the $H_{f_i}(x_0,\xi_0)$ on the $d_\xi f_i(x_0,\xi_0)$, hence is surjective: $T_{(x_0,\xi_0)}(L) \to T_{x_0}(X)$. So, in view of the "if" part of Proposition 3.3.2, locally L is the graph of $d\phi$, ϕ a C^r function on a neighborhood of x_0. ϕ is locally uniquely determined modulo a constant, so we can arrange that $\phi(x_0) = \psi(x_0)$. But $d\phi\big|_Q = d\psi\big|_Q$ then implies (3.6.8). $\square$

The images in X under π of the characteristic strips are called the (*base*) *characteristics* of the functions $f_1,\ldots,f_k$, or rather the manifold V. In general the characteristics through $x \in X$ depend on the choice of $\xi \in T_x(X)^*$. The only possibility that they do not depend on ξ is: $f_i(x,\xi) = \langle v_i(x),\xi\rangle + a_i(x)$, v_i a vector field in X, which means that we are dealing with linear first-order partial differential equations of the form $v_i\phi + a_i = 0$. The condition of the linear independence of the $d_\xi f_i$ means that the $v_i(x)$, $i = 1,\ldots,k$ are linearly independent. The involutiveness condition is equivalent to the existence of functions λ_ℓ^{ij} such that

(3.6.10) $$[v_i,v_j] = \sum_\ell \lambda_\ell^{ij} v_\ell$$

(3.6.11) $$v_i a_j - v_j a_i = -\sum_\ell \lambda_\ell^{ij} a_\ell.$$

So for the v_j we just get the Frobenius integrability condition, the integral manifolds of $[v_1, \ldots, v_k]$ are the base characteristics.

An important example of a *nonlinear* first-order partial differential equation is the equation describing the *characteristic hypersurfaces of an m-th order linear partial differential operator*

$$(3.6.12) \qquad P\left(x, \frac{\partial}{\partial x}\right) = \sum_{|\alpha| \leq m} a_\alpha(x) \cdot \left(\frac{\partial}{\partial x}\right)^\alpha$$

with complex-valued C^∞ coefficients $a_\alpha(x)$. Then for any $\phi \in C^\infty(X)$ the function $e^{-i\tau\phi} P(e^{i\tau\phi})$ is a polynomial in τ of degree m. The coefficient of τ^m is equal to $p(x, d\phi_x)$ where the *principal symbol* p is defined by

$$(3.6.13) \qquad p(x, \xi) = \sum_{|\alpha|=m} a_\alpha(x) \cdot (i\xi)^\alpha.$$

Now a hypersurface that locally is described by an equation of the form $\phi = $ constant, ϕ a real C^∞ function on X, $d\phi \neq 0$, is called *characteristic for the operator P* if

$$(3.6.14) \qquad p(x, d\phi_x) = 0.$$

This is a nonlinear equation for ϕ as soon as $m > 1$. Theorem 3.6.3 now applies in two cases:

(i) p is real and $d_\xi p(x, \xi) \neq 0$ if $p(x, \xi) = 0$, $\xi \neq 0$,

(ii) p is complex and, if $p(x, \xi) = 0$, $\xi \neq 0$, then:
   (iia) $\{\mathrm{Re}\, p, \mathrm{Im}\, p\} = 0$,
   (iib) $d_\xi \mathrm{Re}\, p$, $d_\xi \mathrm{Im}\, p$ are linearly independent.

In both cases the characteristic strips of the "characteristic variety" $p(x, \xi) = 0$ are called the *bicharacteristic strips* of P; in the first case they are one-dimensional and in the second case, two-dimensional. The projections of the bicharacteristic strips into X are called the *bicharacteristic curves* and *surfaces* of P, respectively.

The homogeneity of p in ξ has an interesting consequence for the solutions ϕ of (3.6.14). First assume that p is real, let $(x(t), \xi(t))$ be a bicharacteristic strip for P. Then

$$\frac{d}{dt}\phi(x(t)) = d\phi_{x(t)} \cdot \frac{dx(t)}{dt} = \xi(t) \cdot d_\xi p(x(t), \xi(t))$$

$$= m\, p(x(t), \xi(t)) = 0.$$

Here we used

(3.6.15) $$d_\xi p(x,\xi) \cdot \xi = m\, p(x,\xi)$$

(Euler's identity), which is equivalent to the condition that $\xi \mapsto p(x,\xi)$ is homogeneous of degree m. This can be seen by differentiating $p(x,\tau\xi)$ with respect to τ.

So ϕ is constant on the bicharacteristic curves. In other words, all characteristic hypersurfaces $\phi = \phi(x_0)$ for P, such that $d\phi_{x_0} = \xi_0$, have the bicharacteristic curve through x_0 with respect to ξ_0 in common. For more information on the role of the solutions of (3.6.14) in asymptotic solutions of partial differential equations, see pages 9 and 10 in the Introduction.

Theorem 3.6.3 has been formulated only locally. Obstructions for the existence of global solutions are:

(i) Nonvanishing $H^1(X,\mathbb{R})$. Indeed, let L be a Lagrangian manifold in $f_1 = \cdots f_k = 0$ such that $\pi\colon L \to X$ is a diffeomorphism. Then L is the graph of a 1-form λ on X, $d\lambda = 0$. For every such one form there exists a function ϕ with $d\phi = \lambda$ if and only if $H^1(X,\mathbb{R}) = 0$.

(ii) $\pi\colon L \to X$ can have bijective differential everywhere, without being bijective (multiple covering).

(iii) The mapping $\pi\colon L \to X$ can have singularities, that is, there can exist $(x,\xi) \in L$ such that $D\pi_{(x,\xi)}\colon T_{(x,\xi)}(L) \to T_x(X)$ is no longer injective. In other words, $T_{(x,\xi)}(L)$ has nonzero intersection with the tangent space $\ker D\pi_{(x,\xi)} \simeq T_x(X)^*$ of the fiber $T_x(X)^*$. This is the most important and irrepairable obstruction: it can for instance happen that L is a neighborhood of (x,ξ) on the graph of $d\phi$, but that its flow-out along the Hamilton field H_f leads to a "turning vertical" of the tangent space of L. This effect is, for instance, responsible for the occurrence of *caustics* in geometrical optics. There is a close connection between the theory of "catastrophies" of Thom [80] and the appearance of caustics.

We finally remark that the main point of the proof of Theorem 3.6.2, namely the fact that the condition of being isotropic is invariant under the action of one-parameter groups of symplectic transformations Φ_2^t, has been stressed by S. Lie. (See Engel and Faber [27]. Another classic on this subject is Carathéodory [13]. These books also give references to the older literature on Hamilton–Jacobi theory.)

3.7. Conic Lagrangian manifolds

A submanifold L of $T^*(X) \setminus 0 = \{(x, \xi) \in T^*(X);\ \xi \neq 0\}$ is called *conic* if

$$(3.7.1) \qquad (x, \xi) \in L \Rightarrow (x, \tau\xi) \in L \qquad \text{for all } \tau > 0.$$

The set of $(x, \tau\xi)$, $\tau > 0$ is called the *cone axis* through (x, ξ). Elements $u \in T_{(x,\xi)}(T^*(X))$ will be denoted by $(\delta x, \delta \xi)$, $\delta x = D\pi_{(x,\xi)}(u) \in T_x(X)$. If $\delta x = 0$ then u is tangent to the fiber and can invariantly be written as $(0, \delta\xi)$, $\delta\xi \in T_x(X)^*$. Note that $\delta\xi = 0$ or $(\delta x, \delta\xi) \mapsto \delta\xi$ do not have coordinate invariant meaning.

Proposition 3.7.1. *An n-dimensional closed submanifold L of $T^*(X) \setminus 0$ is a conic Lagrangian manifold if and only if the canonical 1-form α vanishes on L.*

Proof. If L is conic then $(0, \xi) \in T_{(x,\xi)}(L)$ for all $(x, \xi) \in L$. If L in addition is isotropic, then

$$(3.7.2) \qquad \alpha_{(x,\xi)} \begin{pmatrix} \delta x \\ \delta \xi \end{pmatrix} = \xi\, \delta x = \sigma \left(\begin{pmatrix} 0 \\ \xi \end{pmatrix}, \begin{pmatrix} \delta x \\ \delta \xi \end{pmatrix} \right) = 0$$

for all $\begin{pmatrix} \delta x \\ \delta \xi \end{pmatrix} \in T_{(x,\xi)}(L)$.

If conversely $\alpha = 0$ on L, then $\sigma = d\alpha = 0$ on L, so L is Lagrangian. Moreover (3.7.2) shows that $(0, \xi) \in T_{(x,\xi)}(L)^\sigma = T_{(x,\xi)}(L)$ so the vector field $(x, \xi) \to (0, \xi)$ is tangent to L. The cone axes, which are the integral curves of L, therefore are contained in L. Note that these curves cannot run off L because L is assumed to be closed. $\qquad \square$

If V is a k-dimensional submanifold of X, then we define its *normal bundle* in $T^*(X)$ by:

$$(3.7.3) \quad V^\perp = \{(x, \xi) \in T^*(X);\ x \in V,\ \xi(\delta x) = 0 \text{ for all } \delta x \in T_x(V)\}.$$

It is clear that $\{\xi \in T_x(X)^*;\ \xi = 0 \text{ on } T_x(V)\}$ is an $(n - k)$-dimensional linear subspace of $T_x(X)^*$ for all $x \in V$, so $V^\perp$ is an n-dimensional C^∞ submanifold of $T^*(X)$ (and at the same time a vector bundle over V).

Proposition 3.7.2. *$V^\perp \setminus 0$ is a conic Lagrangian manifold. Conversely, if L is conic Lagrangian and the rank of $D\pi_{(x,\xi)} \colon T_{(x,\xi)}(L) \to T_x(X)$ is* constant equal to k for all $(x, \xi) \in L$, then each $(x, \xi) \in L$ has a conic neighborhood Γ such that

(i) $V = \pi(L \cap \Gamma)$ is a k-dimensional C^∞ submanifold of X,

(ii) $L \cap \Gamma$ is an open subset of $V^\perp$.

Proof. Tangent vectors at (x, ξ) to $V^\perp$ are of the form $(\delta x, \delta \xi)$, $\delta x \in T(V)$. Because $\xi \perp T_x(V)$,

$$\alpha_{(x,\xi)} \begin{pmatrix} \delta x \\ \delta \xi \end{pmatrix} = \xi(\delta x) = 0.$$

For the converse we remark that (i) follows by the implicit function theorem, whereas $\Gamma \cap L \subset V^\perp$ follows from the fact that α vanishes on L. Because $\dim L = \dim V^\perp$, it follows that $\Gamma \cap L$ is open in $V^\perp$. $\qquad\square$

Remark. If L in addition is closed and $n - k = \operatorname{codim} V > 1$, then we can choose Γ such that $\Gamma \cap L = V^\perp$. The reason is that in this case $L \cap T_x(X)^*$ is an open and closed subset of $T_x(V)^\perp \setminus \{0\}$. (If $\operatorname{codim} V = 1$ then $T_x(V)^\perp \setminus \{0\}$ has two components.)

In general the collection of points (x, ξ) where the rank of $D\pi_{(x,\xi)}$: $T_{(x,\xi)}(L) \to T_x(X)$ is $\geq k$, forms an open conic subset of L. The rank therefore is locally constant in an open dense conic subset of L, so "almost everywhere" L is equal to an open cone in a normal bundle. That this rank can jump follows from the following description.

Proposition 3.7.3. *An n-dimensional submanifold L of $T^*(X) \setminus 0$ is a conic Lagrangian manifold if and only if every $(x, \xi) \in L$ has a conic neighborhood Γ such that $L \cap \Gamma = \Lambda_\phi$ for some nondegenerate phase function ϕ. (See (2.3.10) for the definition.)*

Proof. To prove that Λ_ϕ is a conic Lagrangian manifold we must show that $\alpha = 0$ on Λ_ϕ, that is

(3.7.4) $\qquad d_x\phi(\delta x) = 0$ if $d_\theta \phi = 0$, $\qquad (\delta x, \delta \theta) \in T(C_\phi)$.

Now the homogeneity of ϕ implies that $d_\theta \phi \cdot \theta = \phi(x, \theta)$. So $\phi = 0$ on C_ϕ, and therefore

$$d_x\phi(\delta x) + d_\theta\phi(\delta \theta) = 0 \quad \text{for } (\delta x, \delta \theta) \in T(C_\phi),$$

which implies (3.7.4).

For the converse we have to find a nondegenerate phase function such that $L = \Lambda_\phi$ (locally). Now suppose that for some induced local coordinatization the projection

$$(3.7.5) \qquad L \ni (x_1, \ldots, x_n, \xi_1, \ldots, \xi_n) \mapsto (\xi_1, \ldots, \xi_n)$$

has an injective, hence bijective differential. This means that L is locally of the form

$$\{(x(\xi), \xi), \ \xi \in (\mathbb{R}^n)^*\} \quad \text{for some } x(\xi) \colon (\mathbb{R}^n)^* \to \mathbb{R}^n.$$

The condition that L is Lagrangian implies that $dx = 0$, x regarded as a 1-form on $(\mathbb{R}^n)^*$. So locally $x(\xi) = dH(\xi)$ for some C^∞ function $H(\xi)$. Because L is conic, $x(\xi)$ is homogeneous in ξ of degree 0 and we can choose $H(\xi)$ homogeneous of degree 1. Now take

$$(3.7.6) \qquad \phi(x, \xi) = \langle x, \xi \rangle - H(\xi).$$

Then $d_\xi\phi(x,\xi) = 0 \Leftrightarrow x = dH(\xi)$, $d_x\phi(x,\xi) = \xi$, so $\Lambda_\phi = L$ locally. $\qquad\square$

The problem that remains in the above proof is to choose local coordinates making the above construction possible. This is taken care of by the following lemma. Note that for any finite number of Lagrangian subspaces $M_1, \ldots, M_k$ in $T_{(x,\xi)}(T^*(X))$ one can always find a Lagrangian subspace M that is transversal to all the M_i, $i = 1, \ldots, k$. (The $\Lambda^0(E, M_i)$ are dense in $\Lambda(E)$, see Theorem 3.4.7.) In particular we can choose M transversal to both $T_{(x,\xi)}(L)$ and the tangent space of the fiber.

Lemma 3.7.4. *Let $(x^0, \xi^0) \in T^*(X) \setminus 0$, suppose M is a Lagrangian subspace of $T_{(x^0,\xi^0)}(T^*(X))$ that is transversal to the tangent space of the fiber (the latter space also is Lagrangian). Then there is an induced local coordinatization such that M is equal to the tangent space of the manifold $(\xi_1, \ldots, \xi_n) = (\xi_1^0, \ldots, \xi_n^0)$.*

Proof. On arbitrary induced local coordinates M is of the form $\{(\delta x, \delta \xi); \ \delta \xi_i = \sum_j A_{ij} \delta x_j\}$ for some symmetric matrix A_{ij} (see the proof of Theorem 3.4.7). Because $\xi^0 \neq 0$ we can choose A_{ij}^k, such that $\sum_k A_{ij}^k \xi_k^0 = A_{ij}$. Now substitute

$$(3.7.7) \qquad x_k = y_k - \frac{1}{2} \sum_{i,j} A_{ij}^k y_i y_j.$$

Then on the new induced coordinates $(\delta y, \delta \eta)$ is, at $x = y = 0$, $\xi = \xi^0$, given by (see (3.3.1))

$$\delta y_k = \delta x_k,$$

$$\delta \eta_k = \delta \xi_k - \sum_{i,j} A_{ik}^j \, \delta x_i \, \xi_j^0 = \delta \xi_k - \sum_i A_{ik} \, \delta x_i.$$

But this means that M coincides with the space $\delta \eta_k = 0$. $\square$

As we have seen in Lemma 2.3.5, the minimum number N of auxiliary variables $\theta_1, \ldots, \theta_N$ in the phase function $\phi(x, \theta)$ describing L is equal to $\dim[T_{(x,\xi)}(L) \cap T_{(x,\xi)}(\text{fiber})]$. (The condition that L is conic implies that N is at least equal to 1, because L has the cone axis in common with the fiber.) That this dimension, equal to n minus the rank of projection: $L \to X$, can jump can be seen in the following example.

Example. Take in $T^*(R^2) \setminus 0$ the conic Lagrangian manifold L defined by $\phi(x, \xi) = \langle x, \xi \rangle - H(\xi)$, $H(\xi) = \xi_1^3/\xi_2^2$, (ξ_1, ξ_2) in a conic neighborhood of $(0, 1)$. The projection of L in X has a cusp at the origin, exactly there the rank of the projection dropped from 1 to 0.

It is also easily verified that the normal bundles of submanifolds of V of X are precisely those conic Lagrangian manifolds defined by a phase function

(3.7.8) $$\phi(x, \theta) = \sum_{j=1}^{n-k} \phi_j(x) \cdot \theta_j$$

which is *linear in* θ. Here the ϕ_j are such that

(3.7.9) $$V = \{x \in X; \; \phi_1(x) = \cdots = \phi_{n-k}(x) = 0\}.$$

A characterization similar to Proposition 3.7.3 also holds for arbitrary (non-conic) Lagrangian submanifolds of $T^*(X)$, dropping the homogeneity condition for ϕ. However, for Fourier integral operators we only need the homogeneous case, so we do not go further into this. Instead we conclude this section with an application to partial differential equations of the form

(3.7.10) $$f_1(x, \phi(x), d\phi_x) = \cdots = f_k(x, \phi(x), d\phi_x) = 0,$$

that is depending also on $\phi(x)$. The key is the following:

Lemma 3.7.5. *The normal bundle in $T^*(X \times \mathbb{R})$ of the graph $\{(x, \phi(x));$ $x \in X\}$ in $X \times \mathbb{R}$ of the function ϕ, is equal to*

$$(3.7.11) \qquad \{(x, \phi(x), -\tau\, d_x\phi, \tau);\ x \in X,\ \tau \in \mathbb{R}\}.$$

The proof is immediate. It follows that ϕ is a solution of (3.7.10) if and only if the normal bundle of its graph is contained in the conic manifold

$$(3.7.12) \qquad g_1(x, t, \xi, \tau) = \cdots = g_k(x, t, \xi, \tau) = 0,$$

where the g_i are the homogeneous functions

$$(3.7.13) \qquad g_i(x, t, \xi, \tau) = \tau \cdot f_i(x, t, -\xi/\tau)$$

of degree 1, defined on the open cone

$$(3.7.14) \qquad \{(x, t, \xi, \tau) \in T^*(X \times \mathbb{R});\ \tau \neq 0\}.$$

So the problem is now reduced to the question of finding the conic Lagrangian manifolds contained in the conic manifold (3.7.12). Taking in Theorem 3.6.2 the initial manifold L_0 conic, L is automatically conic because the H_{g_i}-flows commute with the multiplications with positive scalars in the fibers. If in addition the projection $L \to X \times \mathbb{R} \to X$ has surjective differential, it follows that the projection of L in $X \times \mathbb{R}$ is locally the graph of a function ϕ, which then automatically solves (3.7.10). This construction is even more geometric than the proof of Theorem 3.6.3, since it avoids the integration procedure at the end of that proof. (This remark is due to S. Lie.) The exact formulation of the analogue of Theorem 3.6.3 is left to the reader.

3.8. Classical mechanics and variational calculus

In this section we give a brief description of the relation between classical mechanics and variational calculus with symplectic differential geometry.

In classical mechanics the starting point is *Newton's law*

$$(3.8.1) \qquad m_i\, d^2 x_i/dt^2 = F_i.$$

Here m_i is the *mass*, x_i is a *position* coordinate, F_i the *force* coordinate, all with respect to the i-th degree of freedom. It is assumed that there exists a smooth function U, called the potential energy, such that

$$(3.8.2) \qquad F_i = -\partial U/\partial x_i.$$

In other words, $F = -dU$, we see that force is a 1-form on position space X, m is a linear map: $X \to X^*$, $X = \mathbb{R}^n$. Introducing $dx_i/dt = v_i$ (velocity) the second-order system (3.8.1) can be reduced to a first-order system on a space of double dimension. However, if we take the slight variant $mv = p$ (impulse or momentum), we can even write this system in the form

(3.8.3)
$$\frac{dx}{dt} = d_p E$$
$$\frac{dp}{dt} = -d_x E$$

where the *total energy* $E(x, p)$ is given by

(3.8.4) $$E = \sum_i \frac{1}{2} m_i v_i^2 + U(x) = \sum_i \frac{1}{2m_i} p_i^2 + U(x).$$

So we obtain a Hamilton system in $X \times X^*$ with the total energy as a Hamilton function.

The symplectic structure has been used in classical mechanics in a number of ways. We mention here:

1°. Invariance of Hamilton systems under canonical transformations (Theorem 3.5.2).

2°. Many specific examples turned out to be integrable, that is functions $f_2, \ldots, f_n$ could be found such that $E, f_2, \ldots, f_n$ are in involution. See the end of Section 3.5.

3°. Liouville's theorem. This plays a role in the investigation of ergodic properties of the flow. See, for instance, Arnol'd–Avez [7].

4°. Analysis of fixed points and periodic points of the H_E-flow Φ^t. If $\Phi^t(m) = m$ then $D\Phi_m^t$ is a symplectic automorphism of $T_m(M)$, so we can apply Theorem 3.4.3 and Corollary 3.4.4. (If $H_E(m) = 0$ then $DH_E(m) = \frac{d}{dt} D\Phi_m^t\big|_{t=0}$ is an infinitesimal symplectic mapping and we can apply Theorem 3.4.5 and Corollary 3.4.6). The absolute value of the eigenvalues of $D\Phi_m^t$ is important for the stability of the periodic point m: a necessary (but not sufficient) condition for stability is that they all have absolute value equal to 1. Corollary 3.4.4 shows a remarkable rigidity of this condition: if $|\lambda| = 1$, $\lambda \neq \pm 1$ and λ has multiplicity 1 then for each slightly perturbed symplectic automorphism the eigenvalue $\tilde{\lambda}$ still satisfies $|\tilde{\lambda}| = 1$. However, this does not imply that stability of the periodic point is preserved under small perturbations of the Hamilton function, the situation is much more subtle than that.

We now turn to variational calculus. Let $t_0, t_1 \in \mathbb{R}$, $a, b \in X$. Let Γ be the set of all C^2 curves $x: [t_0, t_1] \to X$ such that $x(t_0) = a$, $x(t_1) = b$. Γ is a Banach space with norm

$$(3.8.5) \qquad \max\left\{ |x(t)|, \left|\frac{dx}{dt}(t)\right|, \left|\frac{d^2x}{dt^2}(t)\right| ; \ t \in [t_0, t_1] \right\}.$$

For each $x(\cdot) \in \Gamma$ define the integral

$$(3.8.6) \qquad I(x(\cdot)) = \int_{t_0}^{t} L\left(x(t), \frac{dx}{dt}(t) \right) dt.$$

Here L is apparently a function on the tangent bundle $T(X)$ of X, assume that L is smooth. Then $x(\cdot) \to I(x\cdot))$ is a differentiable function on Γ, $x(\cdot)$ is called an *extremal curve* for the integral (3.8.6) if $DI_{x(\cdot)} = 0$. Now

$$I(x(\cdot) + h(\cdot)) - I(x(\cdot)) = \int_{t_0}^{t_1} \left(d_x L \cdot h + d_{\dot{x}} L \cdot \frac{dh}{dt} \right) dt + o(h)$$

$$= \int_{t_0}^{t_1} \left[d_x L - \frac{d}{dt}(d_{\dot{x}}L)\left(x, \frac{dx}{dt}\right) \right] \cdot h \, dt + o(h) \qquad \text{for } h \to 0,$$

using partial integration. So $x(\cdot)$ is extremal if and only if it satisfies

$$(3.8.7) \qquad \frac{d}{dt}(d_{\dot{x}}L)\left(x(t), \frac{dx}{dt}(t) \right) = (d_x L)\left(x(t), \frac{dx}{dt}(t) \right),$$

called the *Euler–Lagrange* equations. Note that this is a second-order system of ordinary differential equations for $x(t)$.

Now consider the mapping

$$(3.8.8) \qquad \Phi: (x, \dot{x}) \to (x, d_{\dot{x}}L(x, \dot{x}))$$

which is a mapping: $T(X) \to T^*(X)$ in a coordinate-invariant way. We have that

$$D\Phi: (\delta x, \delta \dot{x}) \to (\delta x, d_x d_{\dot{x}} L \cdot \delta x + d_{\dot{x}}^2 L \cdot \delta \dot{x})$$

is bijective if and only if $d_{\dot{x}}^2 L(x, \dot{x})$ is a nondegenerate bilinear form. If this is true for all $x, \dot{x}$ then L is called a *regular Lagrange function* for the variational problem, which is the same as saying that Φ, called the *velocity-momentum transformation defined by L*, is a local diffeomorphism.

Now define $A(x, \dot{x}) = d_{\dot{x}}L(x, \dot{x}) \cdot \dot{x}$ (called the *action* of L) and $E(x, \dot{x}) = A(x, \dot{x}) - L(x, \dot{x})$ (called the *energy* of L). Define the corresponding *energy* $\mathcal{E}$ of L on $T^*(X)$ by $\mathcal{E} \circ \Phi = E$. The mapping $L \mapsto \mathcal{E}$ is called the *Legendre transformation*.

The assertion is now that the Euler–Lagrange system (3.8.7) under Φ transforms into the Hamilton system with Hamilton function $\mathcal{E}$ on $T^*(X)$.

Proof. Denote $\xi = d_{\dot{x}}L(x, \dot{x})$. Differentiation of $(x, d_{\dot{x}}L(x, \dot{x}))$ with respect to $\dot{x}$ gives $d_\xi \mathcal{E} \circ d_{\dot{x}}^2 L = d_{\dot{x}}[d_{\dot{x}}L \cdot \dot{x} - L] = d_{\dot{x}}^2 L \cdot \dot{x}$, so $\dot{x} = d_\xi(x, d_{\dot{x}}L(x, \dot{x}))$ in view of the nondegeneracy of $d_{\dot{x}}^2 L$. Next differentiation of $\mathcal{E}(x, d_{\dot{x}}L(x, \dot{x}))$ with respect to x gives $d_x L(x, \dot{x}) = -d_x \mathcal{E}(s, d_{\dot{x}}L(x, \dot{x}))$. Putting in $x = x(t)$, $\dot{x} = \frac{dx}{dt}(t)$ and $\xi(t) = d_{\dot{x}}L(x(t), \frac{dx}{dt}(t))$ we obtain the equivalence between (3.8.7) and

$$\frac{dx}{dt}(t) = d_\xi \mathcal{E}(x(t), \xi(t))$$

$$\frac{d\xi}{dt}(t) = -d_x \mathcal{E}(x(t), \xi(t)). \qquad \square$$

Conversely, every Hamilton system with Hamilton function $\mathcal{E}(x, \xi)$ such that $d_\xi^2 \mathcal{E}$ is nondegenerate corresponds to the Euler–Lagrange system for some L in the above fashion, so we do obtain in this way the "general" Hamilton systems.

Examples. (1) Classical mechanics arises from the variational equation with Lagrange function

(3.8.9) $$L(x, v) = \sum \tfrac{1}{2} m_i v_i^2 - U(x).$$

In the time of its discovery this relation of classical mechanics with "optimal curves" gave rise to many metaphysical speculations.

The equivalence between (3.8.1) and (3.8.7), with L equal to the difference of the kinetic and potential energy, is due to Lagrange [51], 2^e Part., Sect. IV. His point was that, under arbitrary substitutions of variables, the "force" $\frac{d}{dt}(d_{\dot{x}}L) - d_x L$ transforms as a 1-form (not as a tangent vector to the x-space), which is simpler than the transformation rule for the acceleration $d^2 x(t)/dt^2$. Remarkably, he does not discuss the relation between (3.8.7) and the variational problem $DI_{x(\cdot)} = 0$, Euler's treatment of which he had long before improved.

In [51], 2^e Part., Sect. V, § I, Lagrange introduces the 2-form in $T(X)$, which in our notation is equal to $\Phi^*\sigma$, and proves that it is invariant under the flow, defined by (3.8.7). However, the equivalence of (3.8.7) to the Hamilton system of the Legendre transform $\mathcal{E}$ of L is due to Hamilton and Jacobi (cf. [46]), although [51], 2^e Part., Sect. V, § II contains a version of this for the variational equations with respect to a parameter in the potential energy. Hamilton and Jacobi also stress the equivalence between (3.8.7) and $DI_{x(\cdot)} = 0$, and the relation between H_f and the partial differential equation $f(x, d\varphi(x)) = 0$.

(2) Let X be a (pseudo) Riemannian manifold, that is, a smooth manifold with nondegenerate bilinear form $Q_x(\cdot, \cdot)$ on $T_x(X)$ for each $x \in X$, depending smoothly on $x \in X$. Then the *geodesic flow* in $T(X)$ is given by the Euler–Lagrange equations with

$$(3.8.10) \qquad\qquad L(x, \dot{x}) = \tfrac{1}{2}Q_x(\dot{x}, \dot{x}).$$

The velocity-momentum transformation for L is equal to

$$(3.8.11) \qquad\qquad (x, \dot{x}) \to (x, Q_x(\dot{x})),$$

where now Q_x is regarded as a linear mapping: $T_x(X) \to T_x(X)^*$. So under this natural mapping the flow is transformed into a Hamiltonian one defined by the Hamilton function

$$(3.8.12) \qquad (x, \xi) \mapsto \tfrac{1}{2}Q_x(Q_x^{-1}(\xi), Q_x^{-1}(\xi)) = \tfrac{1}{2}\xi(Q_x^{-1}(\xi)).$$

(These considerations generalize a mechanical system without a potential function, the mass is replaced by $Q_x: T_x(X) \to T_x(X)^*$.)

Our presentation of Sections 3.1, 3.2, 3.5 and 3.8 is inspired by the very nice book by Abraham and Marsden [1]. Also see Godbillon [33] and Souriau [78].

Chapter 4

Global Theory of Fourier Integral Operators

4.1. Invariant definition of the principal symbol

In this section we give a more detailed description of the line bundle L that was indicated at the end of Section 2.3. Recall that the manifold $\Lambda = \Lambda_\phi$ (defined in Lemma 2.3.2) is a conic Lagrange manifold in $T^*(X)\backslash 0$, and according to Proposition 3.7.3 it is the general one.

We start with an interpretation of the factor $|\det Q|^{-\frac{1}{2}}$ in (2.3.23). Here

$$Q = Q_\psi = d^2_{(x,\theta)}(\phi - \psi) = \begin{pmatrix} d^2_x\phi - d^2_x\psi & d_\theta d_x\phi \\ d_x d_\theta\phi & d^2_\theta\phi \end{pmatrix}$$

taken at (x_0, θ_0).

The manifold C_ϕ in $X \times R^N$ is defined by the equation $d_\theta\phi = 0$, so its tangent space is equal to $\ker d_{(x,\theta)}d_\theta\phi$. Now in general, if A is a surjective linear mapping from the $(n + N)$-dimensional vector space E to the N-dimensional vector space F then, given a volume vol_E, respectively, vol_F on E, respectively, F, this mapping defines a volume on $\ker A$ by the following formula:

(4.1.1)
$$\mathrm{vol}_{\ker A}(e_1,\ldots,e_n) \cdot \mathrm{vol}_F(Ae_{n+1},\ldots,Ae_{n+N})$$
$$= \mathrm{vol}_E(e_1,\ldots,e_{n+N}) \quad \text{for all} \quad e_1,\ldots,e_{n+N} \in E$$
$$\text{such that} \quad e_1,\ldots,e_n \in \ker A.$$

$\mathrm{vol}_{\ker A}$ is also called the *quotient* of the pullback of vol_E by vol_F under the mapping A.

In this way C_ϕ can be provided with the quotient of the volume in $X \times R^N$ by the pullback of the volume in R^N under the mapping $d_\theta\phi$. The isomorphism T_ϕ carries this volume in C_ϕ to a volume in Λ, called "vol_ϕ."

On the other hand the transversality of $L = T_{(x_0,\xi_0)}(d\psi)$ to both the tangent space M_1 of the fiber and $M_2 = T_{(x_0,\xi_0)}(\Lambda)$ implies that we have a well defined linear projection p_L of $E = T_{(x_0,\xi_0)}(T^*(X))$ onto M_1 along

P. Buser, *Fourier Integral Operators*, Modern Birkhäuser Classics,
DOI 10.1007/978-0-8176-8108-1_5, © Springer Science+Business Media, LLC 2011

L, which is bijective on M_2. Pull-back of the volume ω in the fiber by p_L defines another volume $p_L^*\omega$ on M_2.

Lemma 4.1.1. $\det Q = p_L^*\omega/\mathrm{vol}_\phi$.

Proof. We have

$$p_L\colon \begin{pmatrix} \delta x \\ \delta \xi \end{pmatrix} \to \delta\xi - d_x^2\psi \cdot \delta x$$

so $Q_1 = p_L \circ DT_\phi$ is equal to

$$Q_1\colon \begin{pmatrix} \delta x \\ \delta \theta \end{pmatrix} \to (d_x^2\phi - d_x^2\psi) \cdot \delta x + d_\theta d_x\phi \cdot \delta\theta.$$

The volume $p_L^*\omega/\mathrm{vol}_\phi$ is equal to the quotient of $Q_1^*\omega$ by the volume vol_2 on $T_{(x_0,\theta_0)}(C_\phi) = \ker d_{(x,\theta)}d_\theta\phi$ defined by $Q_2 = d_{(x,\theta)}d_\theta\phi$ as above. Using that $Q = \begin{pmatrix} Q_1 \\ Q_2 \end{pmatrix}$ we have for $e_1, \ldots, e_n \in \ker Q_2$:

$$
\begin{aligned}
Q_1^*\omega(e_1, &\ldots, e_n) \cdot \mathrm{vol}(Q_2 e_{n+1}, \ldots, Q_2 e_{n+N}) \\
&= \omega(Q_1 e_1, \ldots, Q_1 e_n) \cdot \mathrm{vol}(Q_2 e_{n+1}, \ldots, Q_2 e_{n+N}) \\
&= \mathrm{vol}(Q e_1, \ldots, Q e_{n+N}) = \det Q \cdot \mathrm{vol}(e_1, \ldots, e_{n+N}) \\
&= \det Q \cdot \mathrm{vol}_2(e_1, \ldots, e_n) \cdot \mathrm{vol}(Q_2 e_{n+1}, \ldots, Q_2 e_{n+N}).
\end{aligned}
$$

This proves the lemma.   □

Changing from a function ψ to ψ', with both $L = T_{(x_0,\xi_0)}(d\psi)$ and $L' = T_{(x_0,\xi_0)}(d\psi')$ transversal to M_2, where $\psi(x_0) = \psi'(x_0)$, $d\psi(x_0) = d\psi'(x_0) = \xi_0$, therefore amounts to multiplication of the right-hand side in (2.3.23) by a factor

$$(4.1.2) \qquad |p_{L'}^*\omega/p_L^*\omega|^{-\frac{1}{2}} \cdot \exp\frac{\pi i}{4}(\mathrm{sgn}\, Q_{\psi'} - \mathrm{sgn}\, Q_\psi).$$

Let $\mathcal{L}$ be the space of all Lagrange spaces in E that are transversal to both M_1 and M_2, define $\tilde{\Omega}_{\frac{1}{2}}$ as the collection of all $f\colon \mathcal{L} \to \phi$ such that

$$(4.1.3) \qquad f(L') = f(L) \cdot |p_{L'}^*\omega/p_L^*\omega|^{-\frac{1}{2}}.$$

In other words, for every density ν of order $\frac{1}{2}$ on M_1 (see Section 1.1) and $f \in \tilde{\Omega}_{\frac{1}{2}}$, $f(L)p_L^*\nu$ is a density of order $\frac{1}{2}$ on M_2 that does not depend on the choice of L. Now the right-hand side of (2.3.23) also depends multiplicatively on $u(x_0)$. If we regard this as an element of $\Omega_{\frac{1}{2}}(T_{x_0}(X))$ then the mapping $D\pi_{(x_0,\xi_0)}\colon E \to T_{x_0}(X)$ induces a density $\nu = |\omega|^{\frac{1}{2}}/u(x_0)$ of order

$\frac{1}{2}$ on $M_1 = \ker D\pi_{(x_0,\xi_0)}$. Here ω is the canonical volume in $T^*(X)$. Disregarding the factor $\exp\left(\frac{\pi i}{4}\operatorname{sgn} Q\right)$ for a moment we can view the right-hand side of (2.3.23) as a density of order $\frac{1}{2}$ on $T_{(x_0,\xi_0)}(\Lambda)$ in its dependence on ψ and u.

We now concentrate on the factor $\exp\left(\frac{\pi i}{4}\operatorname{sgn} Q\right)$. Because the signature of a nonsingular quadratic form only jumps by an even number, changing from ψ_2 to ψ_1 only leads to multiplication by some integer power of i. We know that this power $(\operatorname{sgn} Q_{\psi_1} - \operatorname{sgn} Q_{\psi_2})/2$ can only depend on M_2 and $L_j = T_{(x_0,\xi_0)}(d\psi_j)$, $j = 1,2$ and not on the choice of the phase function ϕ defining Λ but we now show this in a more direct fashion.

The transversality of L to M_1 and M_2 implies that

$$M_2 = \{(\tilde{A}\cdot\delta\xi, \delta\xi + d_x^2\psi\cdot\tilde{A}\cdot\delta\xi);\ \delta\xi \in M_1\}$$

for some linear mapping $\tilde{A}\colon M_1 \to T_{x_0}(X)$. Indeed, if $A\colon M_1 \to L$ such that $M_2 = \{\delta\xi + A\cdot\delta\xi;\ \delta\xi \in M_1\}$ then $\tilde{A} = D\pi_{(x_0,\xi_0)}\circ A$.

Lemma 4.1.2. We have $\operatorname{sgn} Q = \operatorname{sgn}\tilde{A} + \operatorname{sgn} d_\theta^2\phi(x_0,\theta_0)$.

Proof. The matrix Q^{-1} has the form $\begin{pmatrix} \tilde{A} & B \\ B^* & C \end{pmatrix}$. Indeed, $\delta x = \tilde{A}\cdot\delta\xi$ is given by the condition that

$$(\delta x, \delta\xi + d_x^2\psi\cdot\delta x) = (\delta x, d_x^2\phi\cdot\delta x + d_\theta d_x\phi\cdot\delta\theta)$$

if $d_x d_\theta\phi\cdot\delta x + d_\theta^2\phi\cdot\delta\theta = 0$, that is $Q\begin{pmatrix}\delta x \\ \delta\theta\end{pmatrix} = \begin{pmatrix}\delta\theta \\ 0\end{pmatrix}$. In other words, $\tilde{A}$ is equal to Q^{-1}, restricted to the $\begin{pmatrix}\delta\xi \\ 0\end{pmatrix}$ followed by taking the x-component.

Writing $Q = \begin{pmatrix} R & S \\ S^* & T \end{pmatrix}$ then the lemma states that $\operatorname{sgn} Q = \operatorname{sgn}\tilde{A} + \operatorname{sgn} T$. Slightly more generally we shall prove that if Q is a bijective symmetric linear map: $E \to E^*$ then

(4.1.4) $$\operatorname{sgn} Q = \operatorname{sgn}(Q|_F) + \operatorname{sgn}(Q^{-1}|_{F^0})$$

for any linear subspace F of E. Here

$$F^0 = \{u \in E^*;\ u(f) = 0 \ \text{ for all } \ f \in F\}.$$

This formulation has the advantage of being coordinate invariant.

Let $F^\perp = Q^{-1}(F^0)$ denote the orthogonal complement of F in E with respect to Q, write $N = F \cap F^\perp$. Choose $G \subset F$ such that $F = N \oplus G$ and $H \subset F^\perp$ such that $F^\perp = N \oplus H$.

Then $F^\perp = (F + F^\perp) \cap G^\perp$, $F = (F + F^\perp) \cap H^\perp$, so $N = (F + F^\perp) \cap G^\perp \cap H^\perp$, so $(G + H)^\perp \cap (G + H) = (G^\perp \cap H^\perp) \cap (G + H) = 0$. Writing $K = (G + H)^\perp$ we therefore have $E = G \oplus K \oplus H$, G, K, H being mutually orthogonal.

Now N is an isotropic subspace of K, and $\dim K = \dim E - \dim(G + H) = \dim E - (\dim N^\perp - \dim N) = 2 \cdot \dim N$. Because Q is nondegenerate on K we can find for each basis $e_1, \ldots, e_\ell$ of N a set of vectors $f_1, \ldots, f_\ell$ in K such that $(Qe_i)(f_j) = \delta_{ij}$. Write N' for the span of $f_1, \ldots, f_\ell$, and $K = N \oplus N'$.

So for the splitting $E = G \oplus (N \oplus N') \oplus H$ we now have

$$Q = \begin{pmatrix} T_0 & & \\ & \begin{matrix} 0 & I \\ I & 0 \end{matrix} & \\ & & R_0 \end{pmatrix}, \quad \text{so} \quad Q^{-1} = \begin{pmatrix} T_0^{-1} & & \\ & \begin{matrix} 0 & I \\ I & 0 \end{matrix} & \\ & & R_0^{-1} \end{pmatrix}$$

(blank spaces denote zeros). Because $\operatorname{sgn} \begin{pmatrix} 0 & I \\ I & 0 \end{pmatrix} = 0$ we find $\operatorname{sgn} Q = \operatorname{sgn} T_0 + \operatorname{sgn} R_0 = \operatorname{sgn} T_0 + \operatorname{sgn} R_0^{-1} = \operatorname{sgn} T + \operatorname{sgn} A$. $\qquad\square$

Note that Lemmas 4.1.1 and 4.1.2 not only give the dependence of the right-hand side of (2.3.23) when varying ψ and keeping ϕ fixed, but also when varying ϕ, keeping ψ fixed. This explains the quantity $\operatorname{sgn} d_\theta^2 \tilde{\phi} - \operatorname{sgn} d_\theta^2 \phi$ occurring in Hörmander [40], section (3.2.10).

Now observe that $\sigma \left(\tilde{A} \begin{pmatrix} 0 \\ \delta\xi \end{pmatrix}, \begin{pmatrix} 0 \\ \delta\eta \end{pmatrix} \right) = -\langle A \cdot \delta\xi, \delta\eta \rangle$, so

(4.1.5) $$\operatorname{sgn} A = -\operatorname{sgn}(M_1, L; M_2),$$

where the right-hand side is defined in Definition 3.4.3. In view of Theorem 3.4.12 we find that

(4.1.6) $$(\operatorname{sgn} Q_{\psi_1} - \operatorname{sgn} Q_{\psi_2})/2 = s(M_1, M_2; L_1, L_2)$$

if $L_j = T_{(x_0, \xi_0)}(d\psi_j)$, $j = 1, 2$. The index s was introduced in Definition 3.4.2.

Now let Λ be an arbitrary Lagrange manifold in $T^*(X)$. For each $(x, \xi) \in \Lambda$, $\mathcal{L}(x, \xi)$ is the set of Lagrange subspaces in $T_{(x, \xi)}(T^*(X))$ that are transversal to both $M_1(x, \xi) = $ tangent space of the fiber, and $M_2(x, \xi) = $

$T_{(x,\xi)}(\Lambda)$. $L(x,\xi)$ is the set of all mappings $f\colon \mathcal{L}(x,\xi) \to \mathbb{C}$ such that

$$(4.1.7) \qquad f(L_1) = i^{s(M_1(x,\xi),M_2(x,\xi);L_1,L_2)} \cdot f(L_2),$$

for all $L_1, L_2 \in \mathcal{L}(x,\xi)$. Of course $L(x,\xi)$ is a one-dimensional vector space over $\mathbb{C}$. The $L(x,\xi)$, $(x,\xi) \in \Lambda$ can be pieced together in a smooth way because s is locally constant, so they form a C^∞ complex line bundle L over Λ, called the *Keller–Maslov line bundle* of Λ. (See Keller [48], Maslov [61].) Note that L has structure group $\mathbb{Z}_4$, that is, the transition of one local trivialization to another is described by multiplying with i^k, $k \in \mathbb{Z}$, which is a multiplicative action of $\mathbb{Z}_4$ on $\mathbb{C}$. Summarizing:

Definition 4.1.1. For a Fourier integral A of order m, defined by a nondegenerate phase function ϕ and an amplitude $a \in S_\rho^{m-(N/2)+(n/4)}(X \times \mathbb{R}^N)$, the *principal symbol of order m* is the element in

$$(4.1.8) \qquad S_\rho^{m+\frac{1}{4}n}(\Lambda, \Omega_{\frac{1}{2}} \otimes L)/S_\rho^{m+\frac{1}{4}n+1-2\rho}(\Lambda, \Omega_{\frac{1}{2}} \otimes L),$$

given by

$$(4.1.9) \qquad \Lambda \ni \alpha \mapsto e^{i\psi(\pi(\alpha),\alpha)} \langle u\, e^{-i\psi(x,\alpha)}, A \rangle.$$

Here $\Lambda = \Lambda_\phi$ is the conic Lagrange manifold in $T^*(X) \setminus 0$ defined by ϕ. $S_\rho^\mu(\Lambda, \Omega_{\frac{1}{2}} \otimes L)$ denotes the symbol space of sections of the complex line bundle $\Omega_{\frac{1}{2}} \otimes L$ over Λ, of growth order μ. Moreover, $u \in C_0^\infty(X, \Omega_{\frac{1}{2}})$ and $\psi \in C^\infty(X \times \Lambda)$, $\psi(x,\alpha)$ is homogeneous of degree 1 in α and the graph of $x \mapsto d_x\psi(x,\alpha)$ intersects Λ transversally at α. Regarding (4.1.9) as a function of such u and ψ, it becomes an element of (4.1.8) as explained above.

We get $\mu = m + n/4$ because $a(x(\alpha), \theta(\alpha)) \cdot |\det Q(\alpha)|^{-\frac{1}{2}}$ has growth order $m - n/4$ and the unit density of order $\frac{1}{2}$ on the fiber of $T^*(X)$ has growth order $n/2$, which has to be added if (4.1.9) is regarded as a density of order $\frac{1}{2}$ on Λ.

For pseudodifferential operators we have a natural trivialization of $\Omega_{\frac{1}{2}} \otimes L$ by taking the principal symbol of the identity as the unit section, making it coincide with the definition of the principal symbol in Section 2.5.

Lemma 4.1.3. *The Keller–Maslov line bundle L over an arbitrary Lagrange manifold Λ in $T^*(X)$ can always be trivialized.*

Proof. This is true in general for a complex line bundle L over a manifold M having local trivializations $\phi_j \colon L|_{U_j} \to U_j \times \mathbb{C}$ such that

$$(4.1.10) \qquad\qquad \phi_k = e^{\psi_{kj} \circ p} \cdot \phi_j$$

for $\psi_{kj} \in C^\infty(U_j \cap U_k)$ satisfying the coboundary condition $\psi_{\ell j} = \psi_{\ell k} + \psi_{kj}$ on $U_j \cap U_k \cap U_\ell$. Here U_j, $j \in J$ denotes an open covering of M and p is the projection: $L \to M$. (In our case the transition functions ψ_{kj} in the exponent are of the form $\frac{\pi i}{2} s_{kj}$, $s_{kj} \in \mathbb{Z}$.) In other words, the ψ_{kj} define an element of $H^2(M, C^\infty)$. However $H^2(M, C^\infty) = 0$ because the cohomology of a fine sheaf is trivial. So there exist $\chi_j \in C^\infty(U_j)$ such that $\psi_{kj} = \chi_k - \chi_j$ on $U_j \cap U_k$. It follows that the functions

$$(4.9.11) \qquad\qquad \tilde{\phi}_j = e^{-\chi_j \circ p} \cdot \phi_j$$

define a global section of L, making L trivial. $\qquad\qquad\qquad\qquad\square$

If $M = \Lambda$ is a conic Lagrange manifold and $L =$ Keller–Maslov line bundle on Λ, then the functions χ_j in (4.1.11) can be chosen homogeneous of degree 0. (Replace Λ by the manifold of cone axes and apply the construction (4.1.11) on this manifold.) In this way $S_\rho^\mu(\Lambda, \Omega_{\frac{1}{2}} \otimes L)$ can be identified with $S_\rho^\mu(\Lambda, \Omega_{\frac{1}{2}})$. With a partition of unity we can also choose some strictly positive density of order $\frac{1}{2}$, homogeneous of degree 0 on Λ, leading to a further identification of $S_\rho^\mu(\Lambda, \Omega_{\frac{1}{2}})$ with $S_\rho^\mu(\Lambda)$. Such identifications, although of a very arbitrary character and not at all reflecting the "true" character of the line bundle $\Omega_{\frac{1}{2}} \otimes L$, will often be used in proofs to reduce statements about sections of $\Omega_{\frac{1}{2}} \otimes L$ to statements about complex-valued functions on Λ.

More natural trivializations of L can be made if, for instance, the projection $\pi \colon \Lambda \to X$ has constant rank. In this case $\operatorname{sgn}(M_1(x,\xi), M_2(x,\xi); L)$ is locally constant as a function of $(x,\xi) \in \Lambda$ and the mapping $L(x,\xi) \to \mathbb{C}$ assigning to $f \in L(x,\xi)$ the number

$$f(L) \cdot \exp\left(\frac{\pi i}{4} \operatorname{sgn}\big(M_1(x,\xi), M_2(x,\xi), L\big) \right),$$

which is independent of $L \in \mathcal{L}(x,\xi)$, leads to a natural trivialization of L. In particular this holds true for normal bundles Λ of submanifolds in X (see Proposition 3.7.2). For pseudodifferential operators, where $\Lambda =$ normal bundle of the diagonal in $X \times X$, the trivialization is compatible with the trivialization of $\Omega_{\frac{1}{2}} \otimes L$ discussed after Definition 4.1.1.

4.2. Global theory of Fourier integral operators

Definition 4.2.1. Let X be an n-dimensional smooth manifold, Λ an immersed conic Lagrange manifold in $T^*(X) \setminus 0$. A *Fourier integral of order m and type ρ defined by* Λ is a distribution $A \in \mathcal{D}'(X, \Omega_{\frac{1}{2}})$ such that

$$(4.2.1) \qquad\qquad A = \sum_{j \in J} A_j,$$

$A_j \in \mathcal{D}'(X, \Omega_{\frac{1}{2}})$ with locally finite supp A_j, $j \in J$, and A_j is a Fourier integral defined by a nondegenerate phase function ϕ_j on an open cone Γ_j in $X \times \mathbb{R}^{N_j}$ such that $(x, \theta) \mapsto (x, d_x\phi_j(x, \theta))$ is a diffeomorphism from $C_{\phi_j} = \{(x, \theta) \in \Gamma_j; \ d_\theta\phi_j(x, \theta) = 0\}$ onto an open cone in Λ. For the amplitude a_j it is required that $a_j \in S^{m-(N_j/2)+(n/4)}(X \times \mathbb{R}^{N_j})$, cone supp $a_j \subset \Gamma_j$.

The space of all such Fourier integrals will be denoted by $I_\rho^m(X, \Lambda)$. Definition 4.1.1 of the principal symbol can be repeated for $A \in I_\rho^m(x, \Lambda)$ with practically no change, leading to an element of

$$S_\rho^{m+\frac{n}{4}}(\Lambda, \Omega_{\frac{1}{2}} \otimes L) / S_\rho^{m+\frac{n}{4}+1-2\rho}(\Lambda, \Omega_{\frac{1}{2}} \otimes L).$$

Note that the principal symbol of A is equal to the sum of principal symbols of the A_j, where the latter is a locally finite sum because the supports form a locally finite system (even after projection into X!).

Theorem 4.2.1. *If the immersion* $\Lambda \to T^*(X) \setminus 0$ *is proper and injective (that is an embedding), then the mapping* $A \mapsto$ *principal symbol of A is an isomorphism:*

$$(4.2.2) \qquad \begin{aligned} &I_\rho^m(X, \Lambda)/I_\rho^{m+1-2\rho}(X, \Lambda) \\ &\to S_\rho^{m+\frac{n}{4}}(\Lambda, \Omega_{\frac{1}{2}} \otimes L)/S_\rho^{m+\frac{n}{4}+1-2\rho}(\Lambda, \Omega_{\frac{1}{2}} \otimes L). \end{aligned}$$

Proof. We will construct a two-sided inverse. Let $S^*(X)$ be the cotangential sphere bundle; we have the projections

$$T^*(X) \setminus 0 \xrightarrow{\pi_S} S^*(X) \xrightarrow{\pi^S} X, \qquad \pi^S \circ \pi_S = \pi,$$

π^S is a proper mapping (that is, pre-images of compact subsets in X are compact in $S^*(X)$.).

The closed conic submanifold Λ of $T^*(X) \setminus 0$ is mapped under π_S onto a closed C^∞ submanifold of $S^*(X)$. Let V_j, $j \in J$ be a locally finite covering of $\pi_S(\Lambda)$ such that $\Lambda_j = \pi_S^{-1}(V_j) = \Lambda_{\phi_j}$ for a nondegenerate phase function ϕ_j for all $j \in J$. Let $\chi_j \in C_0^\infty(\pi_S(\Lambda))$ be a partition of unity on $\pi_S(\Lambda)$ such that supp $\chi_j \subset V_j$.

Let

$$a \in S_\rho^{m+\frac{n}{4}}(\Lambda, \Omega_{\frac{1}{2}} \otimes L)/S_\rho^{m+\frac{n}{4}+1-2\rho}(\Lambda, \Omega_{\frac{1}{2}} \otimes L)$$

be given. Then there exists $A_j \in I_\rho^m(X, \Lambda)$ defined by the phase function ϕ_j and some suitable amplitude $a_j \in S^{n-(N_j/2)+(n/4)}(X \times \mathbb{R}^{N_j})$ such that the principal symbol of A_j is equal to $(\chi_j \circ \pi_S) \cdot a$. Moreover we can choose supp $A_j \subset \pi^S(V_j)$. Because π^S is proper, the $\pi^S(V_j)$ form a locally finite system in X, so $A = \sum A_j$ defines an element of $I_\rho^m(X, \Lambda)$, and its principal symbol is equal to a.

The above construction gives a right inverse, it suffices now to show that this right inverse is surjective. Any $A \in I_\rho^m(X, \Lambda)$ is of the form $\sum_{k \in K} \tilde{A}_k$ with supp $\tilde{A}_k$, $k \in K$ locally finite and $\tilde{A}_k$ defined by a phase function $\tilde{\phi}_k$. Writing the $\tilde{A}_k$ as a sum of Fourier integrals defined by the phase function $\tilde{\phi}_k$ and amplitudes with sufficiently small conic supports and applying Theorem 2.3.4, we see that A can be rewritten as $\sum_{j \in J} A_j$, A_j defined by ϕ_j, as above. $\qquad\square$

We now turn to the global theory of Fourier integral operators $A: C_0^\infty(Y, \Omega_{\frac{1}{2}}) \to \mathcal{D}'(X, \Omega_{\frac{1}{2}})$, that are operators of which the distribution kernels $K_A \in \mathcal{D}'(X \times Y, \Omega_{\frac{1}{2}})$ are elements of $I_\rho^m(X \times Y, \Lambda)$, for some conic Lagrange manifold Λ in $T^*(X \times Y) \setminus 0$. As we have seen in Sections 1.4 and 2.4, an important part is played by the wave front relation $WF'(A)$, which is a closed conic subset of

$$(4.2.3) \quad C = \Lambda' = \{((x, \xi), (y, \eta)) \in T^*(X) \times T^*(Y); \ (x, y, \xi, -\eta) \in \Lambda\}.$$

If we denote by σ_X and σ_Y the canonical 2-form in $T^*(X)$ and $T^*(Y)$, respectively (see Sections 3.3, 3.5), then the property that Λ is a Lagrange manifold in $T^*(X \times Y)$ for $\sigma_{X \times Y}$ is reflected by saying that C is a Lagrange manifold in $T^*(X) \times T^*(Y)$ for $\sigma_X \otimes (-\sigma_Y)$. If C is the graph of a smooth mapping Φ from an open subset of $T^*(Y)$ to $T^*(X)$ then this means that Φ is a canonical transformation: $T^*(Y) \to T^*(X)$ (and necessarily dim $X =$ dim Y). Because C is conic, Φ is homogeneous of degree 1, that is, it commutes with multiplications with positive constants in the fibers.

With this background a conic Lagrange manifold C in $T^*(X \times Y) \setminus 0$ for $\sigma_X \oplus -\sigma_Y$ will be called a *homogeneous canonical relation from* $T^*(Y)$

to $T^*(X)$. The space of operators A with distribution kernel $K_A \in I_\rho^m(X \times Y, \Lambda)$ will also be called the class of *Fourier integral operators defined by the canonical relation* $C = \Lambda'$, and denoted by $I_\rho^m(X, Y; C)$. The mapping $(x, y, \xi, \eta) \to ((x, \xi), (y, -\eta))$ transforms the line bundle $\Omega_{\frac{1}{2}} \otimes L$ over Λ into a line bundle over $C = \Lambda'$ denoted by $\Omega_{\frac{1}{2}} \otimes L_C$. The principal symbol of $A \in I_\rho^m(X, Y; C)$ will be regarded as an element of $S_\rho^{m+(n_X+n_Y)/4}(C, \Omega_{\frac{1}{2}} \otimes L_C)$ modulo the same symbol space of order $m + (n_X + n_Y)/4 - (2\rho - 1)$.

The global form of Theorem 2.4.1 now reads:

Theorem 4.2.2. *Let X, Y, Z be smooth manifolds, $A_1 \in I_\rho^{m_1}(X, Y; C_1)$, $A_2 \in I_\rho^{m_2}(Y, Z; C_2)$, where C_1 and C_2 are homogeneous canonical relations from $T^*(Y)$ to $T^*(X)$ and from $T^*(Z)$ to $T^*(Y)$, respectively. Suppose that $\rho > \frac{1}{2}$ and that the following conditions are satisfied:*

(4.2.4)   *The projection from* $(\text{supp } K_{A_1} \times K_{A_2}) \cap$
            $(X \times (\text{diag } Y) \times Z)$ *into $X \times Z$ is proper,*

(4.2.5) $\eta \neq 0$ *if* $((x, \xi), (y, \eta)) \in C_1$ *or* $((y, \eta), (z, \zeta)) \in C_2$,

(4.2.6) $C_1 \circ C_2 \subset (T^*(X) \times T^*(Z)) \setminus 0$,

(4.2.7) $C_1 \times C_2$ *intersects* $T^*(X) \times (\text{diag } T^*(Y)) \times T^*(Z)$ *transversally.*

Then $C_1 \circ C_2$ is a homogeneous canonical relation from $T^*(Z)$ to $T^*(X)$, the product of A_1 and A_2 is well defined and satisfies

(4.2.8) $A_1 \circ A_2 \in I_\rho^{m_1+m_2}(X, Z; C_1 \circ C_2)$.

From the product formula (2.4.13) it follows immediately that for every (x, ξ), (y, η), (z, ζ) such that $\rho_1 = ((x, \xi), (y, \eta)) \in C_1$ and $\rho_2 = ((y, \eta), (z, \zeta)) \in C_2$ there exists a bilinear map:

(4.2.9) $(\alpha_1, \alpha_2) \mapsto \alpha_1 \cdot \alpha_2 \colon (\Omega_{\frac{1}{2}} \otimes L_{C_1})(\rho_1) \times (\Omega_{\frac{1}{2}} \otimes L_{C_2})(\rho_2)$
            $\to (\Omega_{\frac{1}{2}} \otimes L_{C_1 \circ C_2})(\rho)$

such that the principal symbol of $A_1 \circ A_2$ at $\rho = ((x, \xi), (z, \zeta)) \in C_1 \circ C_2$ is given by

(4.2.10) $\alpha(\rho) = \sum_{\substack{(y,\eta) \text{ such that } ((x,\xi),(y,\eta)) \in C_1, \\ ((y,\eta),(z,\zeta)) \in C_2}} \alpha_1((x, \xi), (y, \eta)) \cdot \alpha_2((y, \eta), (z, \zeta))$.

Here α_1 and α_2 are the principal symbols of A_1 and A_2, respectively. The sum is locally finite because of (4.2.4) and Definition 4.2.1.

In order to find an explicit formula for the product in (4.2.9) we must, according to Definition 4.1.1, test the distribution kernel of $A_1 \circ A_2$ by rapidly oscillating functions. (In Hörmander [40], Section 4.2, the principal symbol of the product is computed in terms of amplitudes and phase functions; it is interesting to compare both methods.) If $A \in I^m(X, Y; C)$ then the simplest way of testing its distribution kernel is by functions of the form $u(x)\, e^{-it\langle x, \xi_0\rangle} \cdot v(y)\, e^{it\langle y, \eta_0\rangle}$, of course provided that the manifold

$$\{((x,\xi),(y,\eta)); \ \xi = \xi_0, \ \eta = \eta_0\}$$

intersects C transversally. The possibility of such a procedure follows from

Lemma 4.2.3. *Let C be a canonical relation from $T^*(Y)$ to $T^*(X)$, $\alpha = (\beta, \gamma) \in C$. Then*

(a) *The set $\tilde{\Lambda}$ of Lagrange spaces L in $E = T_\beta(T^*(X))$ that are transversal to the tangent space M_1 of the fiber and to $\{e \in E, (e, 0) \in T_\alpha(C)\}$ is open and dense in $\Lambda(E)$;*

(b) *For any $L \in \tilde{\Lambda}$ the set of Lagrange spaces L' in $F = T_\gamma(T^*(Y))$ transversal to the tangent space M_1' of the fiber in $T^*(Y)$ and such that $L \times L'$ is transversal to $T_\alpha(C)$, is open and dense in $\Lambda(F)$.*

Proof. Write $\sigma = \sigma_{T^*(X)}$ at β, $\tau = \sigma_{T^*(Y)}$ at $\gamma \cdot A = T_\alpha(C)$ is a Lagrange space in $(E \times F, \sigma \oplus -\tau)$, write dom $A = \{e \in E; \ (e, f) \in A$ for some $f \in F\}$ and $A^{-1}(0) = \{e \in E; \ (e, 0) \in A\}$. Then

(4.2.11) $(\text{dom } A)^\sigma = A^{-1}(0) \subset \text{dom } A,$

Indeed, $\sigma(e, e') = 0$ for all $e' \in \text{dom } A$ implies that $(\sigma \oplus -\tau)((e, 0), (e', f')) = 0$ for all $(e', f') \in A$, hence $(e, 0) \in A^{\sigma \oplus -\tau} = A$. And conversely.

Let M be a Lagrange space in (E, σ) such that $A^{-1}(0) \subset M \subset \text{dom } A$ (see Proposition 3.4.1). Then the denseness of $\Lambda^0(E, M_1) \cap \Lambda^0(E, M)$ in $\Lambda(E)$ (see Theorem 3.4.7) proves (a).

Now let L be a Lagrange space in (E, σ) such that $L \cap A^{-1}(0) = (0)$. Then

$$A(L) = \{f \in F; \ (e, f) \in A \text{ for some } e \in L\}$$

is a Lagrange space in (F, τ). Indeed, if $(e, f) \in A$, $(e', f') \in A$, $e, e' \in L$, then $\tau(f, f') = 0$ because $(\sigma \oplus -\tau)((e, f), (e', f')) = 0$ and $\sigma(e, e') = 0$, so $A(L)$ is isotropic. Let dim $E = 2m$, dim dom $A = 2m - k$, dim $F = 2r$. Then $\dim(L \cap \text{dom } A) = m + (2m - k) - 2m = m - k$ because $L + \text{dom } A = L + A^{-1}(0)^\sigma = (L \cap A^{-1}(0))^\sigma = E$. On the other hand, $\dim A(0) = \dim((0) \times F) \cap A = 2n + (m+n) - \dim(\text{dom } A \times F) = n - m + k$. So $\dim A(L) = (m - k) + (n - m + k) = n$.

$L \times L'$ is transversal to A if and only if $L' \cap A(L) = (0)$, so (b) follows from the denseness of $\Lambda^0(F, M_1') \cap \Lambda^0_s(F, A(L))$ in $\Lambda(F)$. $\qquad\square$

Repeated application of Lemma 4.2.3 and the proof of Proposition 3.7.3 leads to the existence of induced local coordinates in $T^*(X)$, $T^*(Y)$, $T^*(Z)$ such that

$$C_1 = \left\{ ((d_\xi H_1(\xi, \eta), \xi), (-d_\eta H_1(\xi, \eta), \eta)) \right\}$$
$$C_2 = \left\{ ((d_\eta H_2(\eta, \zeta), \eta), (-d_\zeta H_2(\eta, \zeta), \zeta)) \right\}$$

for certain functions $H_1(\xi, \eta)$, $H_2(\eta, \zeta)$ that are homogeneous of degree 1. The transversality of $C_1 \times C_2$ to $T^*(X) \times \text{diag } T^*(Y) \times T^*(Z)$ just means that $d_\eta^2(H_1 + H_2)$ is nondegenerate where $d_\eta(H_1 + H_2) = 0$.

Now testing of $K_{A_1 \circ A_2}$ by $e^{-i\langle x, \xi_0 \rangle + i\langle z, \zeta_0 \rangle} u_1(x) u_2(z)$ leads to an integral of the form

$$\iiiint e^{-i\langle x, \xi_0 \rangle} u_1(x) K_{A_1}(x, y) K_{A_2}(y, z) u_2(z) e^{i\langle z, \zeta_0 \rangle} dx \, dy \, dz$$

$$= (2\pi)^{-n_Y} \iiiint e^{-i\langle x, \xi_0 \rangle} u_1(x) K_1(x, y)$$

$$\cdot e^{i\langle y - \tilde{y}, \eta \rangle} K_{A_2}(\tilde{y}, z) u_2(z) e^{i\langle z, \zeta_0 \rangle} dx \, dy \, d\tilde{y} \, d\eta \, dz,$$

which amounts to testing K_{A_1} by $e^{-i\langle x, \xi_0 \rangle} u_1(x) \cdot e^{i\langle y, \eta \rangle}$, K_{A_2} by $e^{-i\langle y, \eta \rangle} \cdot u_2(z) e^{i\langle z, \zeta_0 \rangle}$, taking the product and integrating over η. Also don't forget the factor $(2\pi)^{-n_Y}$ in front. For simplicity we assume from now on that $u_1(x)$, respectively, $u_2(z)$ locally represent fixed unit densities of order $\frac{1}{2}$ in X, respectively, Z, and we also choose a unit density in Y.

The principal part of the testing of K_{A_1} by $e^{-i\langle x, \xi_0 \rangle + i\langle y, \eta \rangle}$ is equal to the principal symbol of A_1 at

$$((d_\xi H_1(\xi_0, \eta), \xi_0), (-d_\eta H(\xi_0, \eta), \eta))$$

applied to the Lagrange space $\delta\xi = \delta\eta = 0$, multiplied by the factor

$$e^{-i(\langle d_\xi H_1(\xi_0, \eta), \xi_0 \rangle + \langle d_\eta H_1(\xi_0, \eta), \eta \rangle)} = e^{-iH_1(\xi_0, \eta)}.$$

Here we use the homogeneity of H_1. Taking the product with the principal part of $\langle e^{-i\langle y,\eta\rangle + i\langle z,\zeta\rangle}, K_{A_2}\rangle$ and integrating over η we obtain a principal part consisting of the product of the principal symbol of A_1 at $((x_0,\xi_0),$ $(y_0,\eta_0))$ applied to the Lagrange space $\delta\xi = \delta\eta = 0$, the principal symbol of A_2 at $((y_0,\eta_0),(z_0,\zeta_0))$ applied to the Lagrange space $\delta\eta = \delta\zeta = 0$, and finally the principal part of the integral

$$(*) \qquad \int e^{-i(H_1(\xi_0,\eta)+H_2(\eta,\zeta_0))}\,d\eta.$$

Here η_0 is such that $y_0 = -d_\eta H_1(\xi_0,\eta_0) = d_\eta H_2(\eta_0,\zeta_0)$ we denote $x_0 = d_\xi H_1(\xi_0,\eta_0)$, $z_0 = -d_\zeta H_2(\eta_0,\zeta_0)$. The expression "principal part" is meant as $\alpha = ((x_0,\xi_0),(z_0,\zeta_0)) \in C_1 \circ C_2$ goes to infinity along the cone axes, the points y_0, η_0 depend on α.

Writing $Q = -(d_\eta^2 H_1(\xi_0,\eta_0) + d_\eta^2 H_2(\eta_0,\zeta_0))$, the principal part of $(*)$ is equal to

$$(2\pi)^{n_Y/2} \cdot |\det Q|^{-\frac{1}{2}} \cdot \exp\left(\frac{\pi i}{4}\,\mathrm{sgn}\,Q\right),$$

according to the method of stationary phase. We end up with the conclusion that the principal symbol of $A_1 \circ A_2$ at $((x_0,\xi_0),(z_0,\zeta_0))$ applied to the Lagrange space $\delta\xi = \delta\zeta = 0$ is equal to the product of the principal symbol of A_1 at $((x_0,\xi_0),(y_0,\eta_0))$ applied to $\delta\xi = \delta\eta = 0$, the principal symbol of A_2 at $((y_0,\eta_0),(z_0,\zeta_0))$ applied to $\delta\eta = \delta\zeta = 0$ and finally the factor $(2\pi)^{-n_Y} \cdot |\det Q|^{-\frac{1}{2}} \cdot \exp\left(\frac{\pi i}{4}\,\mathrm{sgn}\,Q\right)$. In coordinate free formulation we therefore have obtained:

Proposition 4.2.4. *The product in* (4.2.9) *is given by:*

$$(4.2.12) \qquad (\alpha_1 \cdot \alpha_2)(L_1 \times L_2) = \alpha_1(L_1 \times L) \cdot \alpha_2(L \times L_2)$$
$$\cdot (2\pi)^{-n_Y/2} |\det Q|^{-\frac{1}{2}} \cdot \exp\left(\frac{\pi i}{4}\,\mathrm{sgn}\,Q\right).$$

Here L_1, L, and L_2 are Lagrange spaces in $T_{(x,\xi)}(T^*(X))$, $T_{(y,\eta)}(T^*(Y))$, and $T_{(z,\zeta)}(T^*(Z))$ that are transversal to the tangent spaces of the fiber and such that $L_1 \times L$, $L \times L_2$, and $L_1 \times L_2$ is transversal to $T_{\rho_1}(C_1)$, $T_{\rho_2}(C_2)$, and $T_\rho(C_1 \circ C_2)$, respectively. Q is the quadratic form $Q_1 - Q_2$, where Q_1 and Q_2 are the quadratic forms on the tangent space M_1 of the fiber in $T^*(Y)$ obtained by representing $(T_{\rho_1}C_1)(L_1)$ and $(T_{\rho_2}C_2^{-1})(L_2)$, respectively, in the form $\{x + Ax;\ x \in M_1\}$, $A: M_1 \to L$ as in Definition 3.4.3.

An interesting special class of Fourier integral operators occurs when C is locally the graph of a canonical transformation, that is, $\dim X =$

dim $Y = n$. Then C carries a natural volume ω_C, equal to the pullback of the canonical volume in $T^*(X)$ under the projection: $C \to T^*(X)$. ω_C is also equal to the pullback of the canonical volume in $T^*(Y)$ under the projection: $C \to T^*(Y)$ because C is a Lagrange manifold for $\sigma_{T^*(X)} - \sigma_{T^*(Y)}$ so the pullbacks of the symplectic forms in $T^*(X)$ and $T^*(Y)$ to C coincide. Using the identification

$$S_\rho^{m+n/2}(C, \Omega_{\frac{1}{2}} \otimes L_C) \ni a \mapsto a \cdot |\omega_C|^{-\frac{1}{2}} \in S_\rho^m(C, L_C)$$

the principal symbol of $A \in I_\rho^m(X, Y; C)$ then can be regarded as an element of $S_\rho^m(C, L_C)$.

The class of *pseudodifferential operators of order m on a manifold X* is defined as $L_\rho^m(X) := I_\rho^m(X, X; I)$, I = graph of the identity: $T^*(X) \setminus 0 \to T^*(X) \setminus 0$ =diagonal in $T^*(X) \setminus 0 \times T^*(X) \setminus 0$. In this case the line bundle $\Omega_{\frac{1}{2}} \otimes L$ over I is trivialized by taking the principal symbol of the identity on $\mathcal{D}'(X)$ as the unit section. The identification of I with $T^*(X) \setminus 0$ by the projection onto the first or the second factor then leads to the principal symbol of a pseudodifferential operator as an element of $S_\rho^m(T^*(X) \setminus 0)$. If $X = \mathbb{R}^n$ then this coincides with the principal symbol defined in Section 2.5.

We conclude this section by an immediate extension of the usual elliptic theory of pseudodifferential operators. Let $A \in I_\rho^m(X, Y; C)$ have principal symbol $a \in S_\rho^m(C, L_C)$. Then $c \in C$ is called a *noncharacteristic* point for A if $a = a_0 \cdot s_0$ where $a_0 \in S_\rho^m(C, L_C)$ is homogeneous of degree m and nowhere vanishing, and $|s_0|$ is bounded away from 0 in a conic neighborhood of c. Note that if a is homogeneous of degree m then the characteristic points of A are exactly the zeros of a. A is called *elliptic* if it has no characteristic points.

Theorem 4.2.5. *Let $A \in I_\rho^m(X, Y; C)$ be an elliptic Fourier integral operator of order m ($\rho > \frac{1}{2}$) defined by a bijective homogeneous canonical transformation C from an open conic subset Γ of $T^*(Y) \setminus 0$ into $T^*(X) \setminus 0$. Then for any conic subset Γ_0 of Γ such that Γ_0, respectively, $C(\Gamma_0)$ is closed in $T^*(Y) \setminus 0$, respectively, $T^*(X) \setminus 0$ one can find a properly supported Fourier integral operator $B \in I_\rho^{-m}(Y, X; C^{-1})$ such that*

$$WF'(BA - I_Y) \cap \Gamma_0 = \emptyset \quad \text{and} \quad WF'(AB - I_X) \cap C(\Gamma_0) = \emptyset.$$

Proof. Let $a \in S_\rho^m(C, L_C)$ be a principal symbol for A. The ellipticity of A implies that there exists $b \in S_\rho^{-m}(C^{-1}, L_{C^{-1}})$ such that $b \cdot a = 1 = $ principal symbol of the identity $\in L^0(Y)$, on a conic neighborhood of diag(Γ_0)

(identified with Γ_0). Choosing $b = 0$ outside a sufficiently small conic neighborhood of $\{(C(y, \eta), (y, \eta)); (y, \eta) \in \Gamma_0\}$ and using that the image of Γ_0, respectively, $C(\Gamma_0)$ is compact in $S^*(Y)$, respectively, $S^*(X)$ over compact subsets of Y, respectively, X, we can find a properly supported $B_0 \in I_\rho^{-m}(Y, X; C^{-1})$ with principal symbol equal to b.

So $B_0 A = I + R$, where $R \in L_\rho^0(Y)$ has vanishing principal symbol on a conic neighborhood of Γ_0. Applying the usual recurrent procedure for pseudodifferential operators (see, for instance, the Introduction) we can find a properly supported $K \in L_\rho^0(Y)$ such that $WF(K(I + R) - I) \cap \Gamma_0 = \emptyset$. Taking $B = K B_0 \in I_\rho^{-m}(Y, X; C^{-1})$ we have obtained that $WF'(BA - I_Y) \cap \Gamma_0 = \emptyset$.

By a similar procedure we can find a properly supported $\tilde{B} \in I_\rho^{-m}(Y, X; C^{-1})$ such that $WF'(A\tilde{B} - I_X) \cap C(\Gamma_0) = \emptyset$. Multiplying $BA - I$ to the right by $\tilde{B}$ we see that $WF'(B - \tilde{B}) \cap \{((y, \eta), C(y, \eta)); (y, \eta) \in C_0\} = \emptyset$, and it follows that also $WF'(AB - I_X) \cap C(\Gamma_0) = \emptyset$. $\quad\square$

If C is bijective from $T^*(Y) \setminus 0$ onto $T^*(X) \setminus 0$ then Theorem 4.2.5 implies that we can find a properly supported $B \in I_\rho^{-m}(Y, X; C^{-1})$ such that both $BA - I_Y$ and $AB - I_X$ are integral operators with C^∞ kernel, that is, B is a *two-sided parametrix* for A. However, in a number of applications we need the "microlocal" (= local in the cosphere bundle = in conic neighborhoods in the cotangent bundle) form of Theorem 4.2.5 rather than the global one, the extreme being the case that $\Gamma_0 = $ cone axis through (y, η), A is noncharacteristic at $(C(y, \eta), (y, \eta))$.

4.3. Products with vanishing principal symbol

The principal symbol of a pseudodifferential operator $A \in L^m(X)$ is invariantly defined as an element of $S^m(T^*(X) \setminus 0)/S^{m-1}(T^*(X) \setminus 0)$ (for simplicity we take $\rho = 1$), that is only gives information of the full symbol in local coordinates modulo symbols of order $m - 1$. Under coordinate transformations κ the computation of the full symbol modulo S^{m-k} needs derivatives of κ up to the order k, that is, the full symbol of class S^m/S^{m-k} is only invariantly defined as a function on the k-jet bundle of X (see Weishu Shih [75]). Admitting only volume preserving coordinate transformations, Gårding–Kotake–Leray [32] found an invariant symbol in $S^m(T^*(X))/S^{m-2}(T^*(X))$. However it turns out that this restriction is not needed when working with densities of order $\frac{1}{2}$.

Proposition 4.3.1. *Let $P \in L^m(X)$ be a properly supported pseudo-differential operator: $C^\infty(X, \Omega_{\frac{1}{2}}) \to C^\infty(X, \Omega_{\frac{1}{2}})$. For any $a \in C^\infty(X, \Omega_{\frac{1}{2}})$ and any real valued $\phi \in C^\infty(X)$ we have:*

$$e^{-i\tau\phi(x)} P(e^{i\tau\phi} a)(x) = p(x, \tau \cdot d\phi(x)) \cdot a(x)$$

(4.3.1)

$$+ \frac{1}{i}(\mathcal{L}_v a)(x, \tau) + O(\tau^{m-2}), \quad \tau \to \infty.$$

Here $p \in S^m(T^*(X) \setminus 0)/S^{m-2}(T^*(X) \setminus 0)$ does not depend on ϕ and a, and is called the principal symbol of P modulo S^{m-2}. The vector field $v(x, \tau) = d_\xi p(x, \tau \cdot d\phi(x))$ is regarded as a τ-dependent vector field (of growth order $m-1$) in X, $\mathcal{L}_v = \mathcal{L}_{\mathrm{Re}\,v} + i\mathcal{L}_{\mathrm{Im}\,v}$ is the Lie-derivative along v, acting on half-densities, which for real vector fields is defined as in (3.2.15).

Proof. In local coordinates one has

$$e^{-i\tau\phi(x)} P(e^{i\tau\phi} a)(x) \sim \omega \cdot \sum_\alpha \frac{1}{\alpha!} \left(\frac{1}{i} \frac{\partial}{\partial \xi} \right)^\alpha P(x, \tau \cdot d\phi(x))$$

(4.3.2)

$$\cdot e^{-i\tau\phi(x)} \left(\frac{\partial}{\partial x} \right)^\alpha (e^{i\tau\phi(x)} a_0(x))$$

for $\tau \to \infty$. Here $P(x, \xi)$ denotes the full symbol of P and we have written $a = a_0 \cdot \omega$, $a_0 \in C^\infty(\mathbb{R}^n)$, $\omega = $ unit density of order $\frac{1}{2}$ in $\mathbb{R}^n$. The formula (4.3.2) follows from Taylor development of $e^{i\tau\phi(y)}$ in the formula for $P(e^{i\tau\phi} a)(x)$, followed by partial integrations using

$$e^{i\langle x-y, \xi \rangle}(y - x)^\alpha = \left(-\frac{1}{i} \frac{\partial}{\partial \xi} \right)^\alpha e^{i\langle x-y, \xi \rangle}.$$

Disregarding terms of order τ^{m-2} as $\tau \to \infty$ we obtain

$$e^{-i\tau\phi(x)} P(e^{i\tau\phi} a)(x) = P(x, \tau \cdot d\phi(x)) a(x)$$

$$+ \frac{1}{i} \sum_j \frac{\partial P}{\partial \xi_j}(x, \tau \cdot d\phi(x)) \cdot \frac{\partial a_0}{\partial x_j}(x) \cdot \omega$$

(4.3.3)

$$- \frac{1}{2} \sum_{j,k} \frac{\partial^2 P}{\partial \xi_j \cdot \partial \xi_k}(x, \tau \cdot d\phi(x))$$

$$\cdot i\tau \frac{\partial^2 \phi(x)}{\partial x_j \partial x_k} \cdot a(x) + O(\tau^{m-2}).$$

However, $\mathcal{L}_v \omega = \frac{1}{2}\mathrm{div}\, v \cdot \omega$, so $\mathcal{L}_v a = (va_0) \cdot \omega + \frac{1}{2}\mathrm{div}\, v \cdot a$, which proves (4.3.1) with

$$p(x, \xi) = P(x, \xi) - \frac{1}{2i} \sum_{j=1}^n \frac{\partial^2 P(x, \xi)}{\partial x_j \partial \xi_j}. \qquad \square$$

In most applications the principal symbol P in

$$S^m(T^*(X))/S^{m-1}(T^*(X))$$

is given by a *homogeneous* C^∞ function p_m of degree m on $T^*(X) \setminus 0$. Such a homogeneous function, when it exists, is uniquely determined and is called *the* homogeneous principal symbol of degree m of P. If p denotes the principal symbol of P modulo S^{m-2} defined in Proposition 4.3.1 and P has a homogeneous principal symbol p_m of degree m, then

$$p_{m-1} = p - p_m \in S^{m-1}(T^*(X))/S^{m-2}(T^*(X))$$

will be called the *subprincipal symbol* of order $m - 1$ of P.

Theorem 4.3.2. *Let $P \in L^m(X)$ be a properly supported pseudodifferential operator with homogeneous principal symbol p_m of degree m. Assume that C is a homogeneous canonical relation from $T^*(Y) \setminus 0$ to $T^*(X) \setminus 0$, such that p_m vanishes on the projection of C in $T^*(X) \setminus 0$. If $A \in I_\rho^{m'}(X, Y; C)$ and $a \in S^{m'+(n_X+n_Y)/4}(C, \Omega_{\frac{1}{2}} \otimes L_C)$ is a principal symbol of A, then $PA \in I_\rho^{m+m'-\rho}(X, Y; C)$, and its principal symbol is given by*

(4.3.4) $$\frac{1}{i}\mathcal{L}_{H_{p_m}} a + p_{m-1} \cdot a.$$

Here H_{p_m} is the Hamilton field of p_m lifted to a function on $T^*(X) \setminus 0 \times T^*(Y) \setminus 0$ via the projection onto the first factor. The vector field H_{p_m} is tangent to C so (4.3.4) is well defined. $\mathcal{L}$ denotes Lie-derivative of densities of order $\frac{1}{2}$, the line bundle L is not involved in differentiations because the transition functions are constants. Finally the subprincipal symbol p_{m-1} is pulled back to C under the projection $C \to T^*(X) \setminus 0$ from $C \subset T^*(X) \setminus 0 \times T^*(Y) \setminus 0$ to the first factor.

Proof. Representing A locally by integrals

$$\iint e^{i\phi(x,y,\theta)} a(x, y, \theta) u(y)\, dy\, d\theta$$

we see that the proof follows formally from Proposition 4.3.1 by an application of P under the integral sign. For more details see [22], Section 5.3.

4.4. L^2-continuity

If u and v are densities of order $\frac{1}{2}$ in a manifold X and supp $u \cap$ supp v is compact we write

$$(u, v) = \langle u, \bar{v} \rangle = \int u \, \bar{v}.$$

For $u \in C_0^\infty(X, \Omega_{\frac{1}{2}})$ we have the L^2-norm $|u| = (u, u)^{\frac{1}{2}}$ and for $u \in C^\infty(X, \Omega_{\frac{1}{2}})$ the L^2-seminorms $|u|_\phi = |\phi u|$ where ϕ ranges over $C_0^\infty(X)$. The completion with respect to these seminorms is denoted by $L^2_{\text{loc}}(X) \subset \mathcal{D}'(X, \Omega_{\frac{1}{2}})$, its elements with compact support form a subspace $L^2_{\text{comp}}(X)$.

The adjoint A^* of an operator $A \in I_\rho^m(X, Y; C)$, where C is a homogeneous canonical relation from $T^*(Y)$ to $T^*(X)$, is defined by

$$(Au, v) = (u, A^*v), \quad v \in C_0^\infty(X, \Omega_{\frac{1}{2}}), \quad u \in C_0^\infty(Y, \Omega_{\frac{1}{2}}).$$

If A is locally represented by

$$(Au)(x) = \iint e^{i\phi(x, y, \theta)} a(x, y, \theta) \, u(y) \, dy \, d\theta$$

then

$$(A^*v)(y) = \iint e^{-i\phi(x, y, \theta)} \overline{a(x, y, \theta)} \, v(x) \, dx \, d\theta$$

so we obtain immediately:

Theorem 4.4.1. *Let $A \in I_\rho^m(X, Y; C)$ where C is a homogeneous canonical relation from $T^*(Y)$ to $T^*(X)$, with principal symbol a. Then $A^* \in I_\rho^m(Y, X; C^{-1})$ with principal symbol $s^*\bar{a}$, if s denotes the mapping*

$$((x, \xi), (y, \eta)) \to ((y, \eta), (x, \xi)).$$

Corollary 4.4.2. *If $A \in I_\rho^0(X, Y; C)$ and C is locally in $T^*(X) \times T^*(Y)$ the graph of a canonical transformation $(\dim X = \dim Y)$ then A is continuous: $L^2_{\text{comp}}(Y, \Omega_{\frac{1}{2}}) \to L^2_{\text{loc}}(X, \Omega_{\frac{1}{2}})$.*

Proof. A is a locally finite sum of compactly supported $A_j \in I_\rho^0(X, Y; C_j)$ where C_j is the graph of a bijective homogeneous canonical transformation from an open cone in $T^*(Y) \backslash 0$ to an open cone in $T^*(X) \backslash 0$. From Theorem 4.4.1 it follows that $A_j^* A_j \in I_\rho^0(Y)$, which is L^2-continuous, because of Theorem 2.5.4. This implies that the A_j are L^2-continuous and hence A is. $\qquad\square$

If C is not locally the graph of a canonical transformation then one can split up the x, respectively, y-coordinates in groups (x', x''), respectively, (y', y'') and write

$$(Au)(x) = \int (A_{(x'',y'')}u(\cdot, y''))(x')\, dy'',$$

where $A_{(x'',y'')}$ denotes the Fourier integral operator (locally) defined by the phase function $(x', y', \theta) \to \phi(x', x'', y', y'', \theta)$ and corresponding amplitude, depending on the parameters x'', y''. A straightforward estimate shows that A is continuous: $L^2_{\text{comp}} \to L^2_{\text{loc}}$ if the operators $A_{(x'',y'')}$ from y'-space to x'-space are so with L^2-norms bounded as x'', y'' vary. This is the case if the $A_{(x'',y'')}$ are operators satisfying the conditions of Corollary 4.4.2.

In view of Hörmander [40], pp. 170–172, this procedure works with k-dimensional x'- and y'-spaces if and only if

(4.4.1)   The projections $C \to X$ and $C \to Y$ have surjective differential,

and

(4.4.2)   The rank of the projection: $C \to T^*(X)$, respectively, $C \to T^*(Y)$ is $\geq \dim X + k$, respectively, $\geq \dim Y + k$.

If A has order m then according to (2.4.22) the operator $A_{(x'',y'')}$ has order $m + (n_X + n_Y - 2k)/4$ so A is L^2-continuous if this number is ≤ 0. This is Theorem 4.3.2 in Hörmander [40].

Multiplication of an operator A from the left or the right, by means of a Fourier integral operator of order 0 defined by a canonical transformation, the L^2-continuity properties of A will not change, in view of Corollary 4.4.2. However (4.4.1) can change rather drastically under multiplication of C to the right or the left by canonical transformations (note that (4.4.2) remains unchanged). This can be used to relax the condition (4.4.1) substantially. We begin this with the following preparation:

Lemma 4.4.3. *Let $(x, \xi) \in T^*(X)\backslash 0$, $(\tilde{x}, \tilde{\xi}) \in T^*(\tilde{X})\backslash 0$, $\dim X = \dim \tilde{X}$, and let Q be a symplectic linear mapping: $T_{(x,\xi)}(T^*(X)) \to T_{(\tilde{x},\tilde{\xi})}(T^*(\tilde{X}))$ such that $Q\begin{pmatrix} 0 \\ \xi \end{pmatrix} = \begin{pmatrix} 0 \\ \tilde{\xi} \end{pmatrix}$. Then there exists a homogeneous canonical transformation Φ from a conic neighborhood of (x, ξ) to $T^*(\tilde{X})$ such that $\Phi(x, \xi) = (\tilde{x}, \tilde{\xi})$ and $D\Phi_{(x,\xi)} = Q$.*

Proof. It is sufficient to show that if Z is a manifold, $(z, \zeta) \in T^*(Z) \setminus 0$, L a Lagrange subspace of $T_{(z,\zeta)}(T^*(Z) \setminus 0)$ such that $\begin{pmatrix} 0 \\ \zeta \end{pmatrix} \in L$, then there exists a conic Lagrange submanifold Λ of $T^*(Z) \setminus 0$ through (z, ζ) such that $T_{(z,\zeta)}(\Lambda) = L$. Indeed, applying this to $Z = X \times \tilde{X}$ and using the identification

$$((x, \tilde{x}), (\xi, \tilde{\xi})) \mapsto ((x, \xi), (\tilde{x}, \tilde{\xi})) \colon (T^*(X \times \tilde{X}), \sigma_{X \times \tilde{X}})$$
$$\rightarrow (T^*(X) \times T^*(\tilde{X}), \sigma_X - \sigma_{\tilde{X}})$$

we see that the lemma follows because the tangent space is given by Q and therefore must be the tangent space of the graph of a mapping: $T^*(X) \rightarrow T^*(X)$.

On induced coordinates, conic Lagrange manifolds are of the form

$$\Lambda = \{(dH(\zeta), \zeta); \ \zeta \in \mathbb{R}^{n_z}\}$$

for some homogeneous function H of degree 1, so the problem is reduced to finding such H with prescribed $dH(\zeta_0) = z_0$ and $d^2H(\zeta_0) = B$. Without loss of generality we may assume that $z_0 = 0$.

Let H_S be an arbitrary smooth function on the sphere S of radius $|\zeta_0|$, such that $H_S(\zeta_0) = 0$, $dH_S(\zeta_0) = 0$, $d^2H_S(\zeta_0) = B\big|_{T_{\zeta_0}(S)}$, and let H be the unique homogeneous extension of H. Differentiating Euler's identity $dH(\zeta) \cdot \zeta = H(\zeta)$, we obtain

$$d^2\overset{\cdot}{H}(\zeta) \cdot \zeta = 0.$$

On the other hand, $B \cdot \zeta_0 = 0$ reflects that $L = \{(B \cdot \delta\zeta, \delta\zeta); \ \delta\zeta \in \mathbb{R}^{n_z}\}$ contains $(0, \zeta_0)$, so $d^2H(\zeta_0) = B$ as desired. Also $dH(\zeta_0) = 0$, because $dH(\zeta_0) \cdot \zeta_0 = H(\zeta_0) = 0$ and $dH_S(\zeta_0) = 0$. $\qquad\square$

Now the condition

(4.4.1′)   The differential at $\rho = ((x, \xi), y, \eta))$ of the projection
$C \rightarrow X$ is surjective

apparently means that

(4.4.3) $\dim T_\rho(C) + M_1 = E$

in the notation of Lemma 4.2.3. This is also equivalent to

(4.4.4) $(T_\rho C)^{-1}(0) \cap M_1 = (0),$

where $(T_\rho C)^{-1}(0)$ is isotropic. Now assume that (4.4.4) is weakened to

(4.4.5) $$\begin{pmatrix} 0 \\ \xi \end{pmatrix} \notin (T_\rho\, C)^{-1}(0).$$

Then we can find Lagrange spaces M_0 and L in E, such that $M_0 \cap L = (0)$, $M_1 \cap L = (0)$, $\begin{pmatrix} 0 \\ \xi \end{pmatrix} \in M_0$, and $(T_\rho C)^{-1}(0) \subset L$; see the proof of Theorem 3.4.2 for the construction of such Lagrange subspaces "in general position". Subsequently, we can find a symplectic linear mapping $Q : E \to E$, such that $Q(M_0) = M_1$, $Q(L) = L$ and $Q\begin{pmatrix} 0 \\ \xi \end{pmatrix} = \begin{pmatrix} 0 \\ \xi \end{pmatrix}$.

Multiplying A from the left by an elliptic Fourier integral operator R of order 0 defined by a homogeneous canonical transformation Φ with $\Phi(x,\xi) = (x,\xi)$, $D\Phi_{(x,\xi)} = Q$, we obtain that RA satisfies (4.4.1') if A only satisfies (4.4.5). Moreover, in view of Corollary 4.4.2 and Theorem 4.2.5, the properties of RA with respect to L^2-continuity are the same as those of A, if $WF'(A)$ is contained in a sufficiently small cone around (x,ξ,y,η). Using a similar multiplication to the right we obtain:

Theorem 4.4.4. *Let $C \subset T^*(X)\backslash 0 \times T^*(Y)\backslash 0$ be a homogeneous canonical relation such that*

(i$_a$) $$\left(\begin{pmatrix} 0 \\ \xi \end{pmatrix}, \begin{pmatrix} 0 \\ 0 \end{pmatrix} \right) \notin T_{((x,\xi),(y,\eta))}(C),$$

(i$_b$) $$\left(\begin{pmatrix} 0 \\ 0 \end{pmatrix}, \begin{pmatrix} 0 \\ \eta \end{pmatrix} \right) \notin T_{((x,\xi),(y,\eta))}(C) \ and$$

(ii) *The differentials of the projections $C \to T^*(X)$ and $C \to T^*(Y)$ have rank $\geq \dim X + k$ and $\geq \dim Y + k$, respectively.*

Then every $A \in I_\rho^m(X,Y;C)$ is continuous:

$$L^2_{\mathrm{comp}}(Y,\Omega_{\frac{1}{2}}) \to L^2_{\mathrm{loc}}(X,\Omega_{\frac{1}{2}}),$$

provided that $m + (n_X + n_Y - 2k)/4 \leq 0$.

Define $H^s_{\mathrm{loc}}(X)$ as the space of $u \in \mathcal{D}'(X,\Omega_{\frac{1}{2}})$ such that $Au \in L^2_{\mathrm{loc}}(X)$ for all properly supported $A \in L^s(X)$ or, which amounts to the same, for some elliptic properly supported $A \in L^s(X)$.

Corollary 4.4.5. *Under the conditions* (i), (ii) *in Theorem 4.4.4, every* $A \in I_\rho^m(X, Y; C)$ *is continuous:*

$$H_{\text{comp}}^s(Y) \to H_{\text{loc}}^{s-m-(n_X+n_Y-2k)/4}(X)$$

for each $s, m \in \mathbb{R}$.

Example. Let Φ be a differentiable mapping: $X \to Y$. Then $\Phi^*: C^\infty(Y) \to C^\infty(X)$ is a Fourier integral operator defined by the canonical relation

$$\tilde{\Phi} = \{((x, \xi), (y, \eta)); \ y = \Phi(x), \ \xi = {}^t D\Phi_x \cdot \eta\},$$

and of order $m = \frac{1}{4}(n_Y - n_X)$. See (2.4.4), (2.4.22). It follows easily that

$$C = \tilde{\Phi} \setminus \{((x, 0), (\Phi(x), \eta)); \ {}^t D\Phi_x \cdot \eta = 0\}$$

satisfies (i), (ii) in Theorem 4.4.4 with

$$k = \min\{\text{rank } D\Phi_x; \ x \in X\}.$$

Note that C does not satisfy (4.4.1) unless Φ is a submersion, that is, has a surjective differential everywhere.

Proposition 4.4.6. *Let* Φ *be a* C^∞ *mapping:* $X \to Y$; *write*

$$N = \{(\Phi(x), \eta) \in T^*(Y) \setminus 0; \ x \in X, \ {}^t D\Phi_x \cdot \eta = 0\}.$$

Then, for any $A \in L^0(X)$ *such that* $WF(A) \cap N = \emptyset$, *the operator* $\Phi^* \circ A$ *is continuous:* $H_{\text{comp}}^s(Y) \to H_{\text{loc}}^{s-(1/2)(n_Y-k)}(X)$ *for all* $s \in \mathbb{R}$.

The pseudodifferential operator A is only introduced in Proposition 4.4.6 in order to cut off $WF(Au)$ from N, making it possible to pull $Au \in \mathcal{D}'(Y, \Omega_{\frac{1}{2}})$ back under Φ (see Theorem 1.4.1). Operators similar to $\Phi^* \circ A$ were considered by Sjöstrand [76]. Proposition 4.4.6 can be considered as a variant of the usual continuity properties for restrictions to a submanifold, if $\Phi: X \to Y$ is an embedding. Note that in this case the number $n_Y - k$ is just the codimension of X in Y.

Regarding a conic Lagrange manifold Λ in $T^*(X) \setminus 0$ as a canonical relation: $T^*(X) \to T^*(\text{point})$, we see that Theorem 4.4.4 does not apply to obtain smoothness of elements $u \in I^m(X, \Lambda)$. In fact in this case the result is slightly worse:

Theorem 4.4.7. $I_\rho^m(X, \Lambda) \subset H_{\text{loc}}^s(X)$ *if* $m + n/4 + s < 0$. *Conversely, if* $u \in I_\rho^m(X, \Lambda)$ *has a noncharacteristic point and* $m + n/4 + s \geq 0$, *then* $u \notin H_{\text{loc}}^s(X)$.

Proof. It is sufficient to show that if $u \in I_\rho^m(X, \Lambda)$ and $WF(u)$ is contained in a small conic neighborhood Γ of $(x_0, \xi_0) \in \Lambda$, then $u \in H_{\text{loc}}^0 = L_{\text{loc}}^2$ if $m + n/4 < 0$ and $u \notin L_{\text{loc}}^2$ if $m + n/4 \geq 0$ and (x_0, ξ_0) is a noncharacteristic point for u.

Near (x_0, ξ_0) the manifold Λ is defined by $x = dH(\xi)$ on suitable local coordinates (see the proof of Proposition 3.7.3). It follows that the homogeneous canonical transformation

$$\chi : (x, \xi) \to (x - dH(\xi), \xi)$$

maps Λ into the fiber in $T^*(\mathbb{R}^n) \backslash 0$ over $0 \in R^n$. Choosing $A \in I^0(X, \mathbb{R}^n; \chi)$, $B \in I^0(R^n, X; \chi^{-1})$ such that $\Gamma \cap WF(AB - I) = \emptyset$ according to Theorem 4.2.5, we see that $ABu - u \in C^\infty$, hence $u \in L_{\text{loc}}^2 \Rightarrow Bu \in L_{\text{loc}}^2 \Rightarrow ABu \in L_{\text{loc}}^2 \Rightarrow u \in L_{\text{loc}}^2$ in view of Corollary 4.4.2.

So we have reduced the proof to the case that

$$u(x) = \int e^{i\langle x, \xi \rangle} a(\xi) \, d\xi, \qquad a \in S_\rho^{m - n/4}.$$

We may also take u of compact support, so $u \in L_{\text{loc}}^2 \Leftrightarrow u \in L^2 \Leftrightarrow a \in L^2$ in view of Parseval's equality. It follows that $u \in L_{\text{loc}}^2$ if $2(m - n/4) < -n$ and $u \notin L_{\text{loc}}^2$ if $2(m - n/4) \geq -n$ and u has a noncharacteristic point. $\square$

Chapter 5

Applications

5.1. The Cauchy problem for strictly hyperbolic differential operators with C^∞ coefficients

Let X be an n-dimensional paracompact C^∞ manifold, $P \in L^m(X)$ a properly supported pseudodifferential operator of order m on X with homogeneous C^∞ principal symbol $p_m(x,\xi)$ of degree m on $T^*(X) \setminus 0$. Denote by

$$(5.1.1) \qquad N = \{(x,\xi) \in T^*(X) \setminus 0; \; p_m(x,\xi) = 0\}$$

the set of characteristic points of P. Because p_m is continuous and homogeneous, N is a closed conic subset of $T^*(X) \setminus 0$.

Proposition 5.1.1. $WF(u) \subset WF(Pu) \cup N$ for any $u \in \mathcal{D}'(X)$.

Proof. Suppose $(x,\xi) \notin WF(Pu) \cup N$. According to Theorem 4.2.5 there exists a properly supported $Q \in L^{-m}(X)$ such that $(x,\xi) \notin WF(QP - I)$. Then $u = QPu + (I - QP)u$ leads to $WF(u) \subset WF(Pu) \cup WF((I - QP)u)$, so $(x,\xi) \notin WF(u)$. $\qquad\square$

If X_0 is a submanifold of X of codimension k then the restriction operator $\rho_0 \colon C^\infty(X) \to C^\infty(X_0)$ is a Fourier integral operator of class $I^{k/4}(X_0, X; R_0)$ defined by the homogeneous canonical relation

$$(5.1.2) \qquad \begin{aligned} R_0 = \{((x_0,\xi_0),(x,\xi)) \in T^*(X_0) \times T^*(X) \setminus 0; \\ x_0 = x, \; \xi_0 = \xi|_{T_{x_0}(X)}\} \end{aligned}$$

(see (2.4.4), (2.4.22)). So, in view of Theorem 1.4.1 and Proposition 5.1.1, $\rho_0 Q u$ is well-defined for any properly supported pseudodifferential operator Q if $u \in \mathcal{D}'(X)$ is a solution of $Pu = f$, $WF(f) \cap X_0^{\perp} = \emptyset$ and $N \cap X_0^{\perp} = \emptyset$. Here $X_0^{\perp}$ denotes the normal bundle of X_0 in $T^*(X)$. The manifold X_0 is called *noncharacteristic for P* if $N \cap X_0^{\perp} = \emptyset$. Moreover, if $f \in C^\infty(X)$ and

P. Buser, *Fourier Integral Operators*, Modern Birkhäuser Classics,
DOI 10.1007/978-0-8176-8108-1_6, © Springer Science+Business Media, LLC 2011

X_0 is noncharacteristic for ρ, then $(x_0, \xi_0) \in WF(\rho_0 Qu)$ only if $(x_0, \xi) \in WF(u)$ for some solution $\xi \in (T_{x_0} X)^*$ of the equations

$$(5.1.3) \qquad\qquad p_m(x_0, \xi) = 0, \qquad \xi\big|_{T_{x_0} X_0} = \xi_0.$$

Assume from now on that p_m is real and $k = \text{codim } X_0 = 1$. Then $\ell(x_0, \xi_0) = \{\xi \in (T_{x_0} X)^*; \, \xi\big|_{T_{x_0} X_0} = \xi_0\}$ is a line and we see that the zeros of p_m on this line are simple if and only if $d_\xi p_m(x_0, \xi) \neq 0$ on $(T_{x_0} X_0)^\perp =$ the orthogonal complement in $(T_{x_0} X)^*$ of $T_{x_0} X_0 \subset T_{x_0} X$. In other words, if and only if

$$(5.1.4) \qquad d_\xi p_m(x_0, \xi) \notin T_{x_0} X_0 \quad \text{when} \quad p_m(x_0, \xi) = 0, \ \ \xi \neq 0.$$

Now $d_\xi p_m(x_0, \xi)$ is the velocity vector in $T_{x_0} X$ of the bicharacteristic curve through x_0, which is the projection into X of the bicharacteristic strip through $(x_0, \xi) \in N$. So *the condition of the simplicity of the zeros of (5.1.3) is equivalent to the condition that all bicharacteristic curves are transversal to X_0.*

Condition (5.1.4) implies that X_0 is noncharacteristic for P. Indeed, if $\xi \in (T_{x_0} X_0)^\perp$, $p_m(x_0, \xi) = 0$, $\xi \neq 0$ then $d_\xi p_m(x_0, \xi) \cdot \xi = m \cdot p_m(x_0, \xi) = 0$ because of Euler's identity, hence $d_\xi p_m(x_0, \xi) \in T_{x_0} X_0$ because the latter space is equal to the orthogonal complement of ξ (note that $T_{x_0} X_0$ has codimension 1 in $T_{x_0} X$).

Condition (5.1.4) also implies that (5.1.3) has only finitely many solutions for every $(x_0, \xi_0) \in T^*(X_0)$. Indeed, if $\xi^{(k)}$, $k = 1, 2, \ldots$ is an infinite sequence of different solutions, then $|\xi^{(k)}| \to \infty$ as $k \to \infty$ because the solutions are isolated points. Taking a subsequence if necessary we have $\xi^{(k)} / |\xi^{(k)}| \to \xi$ for some $\xi \in (T_{x_0} X)^*$ with $|\xi| = 1$. Because $\xi^{(k)} / |\xi^{(k)}|\big|_{T_{x_0} X_0} = \xi_0 / |\xi^{(k)}|$ it follows that $\xi\big|_{T_{x_0} X_0} = 0$. On the other hand, $p_m(x_0, \xi^{(k)} / |\xi^{(k)}|) = 0$ because of the homogeneity of p_m, so $p_m(x_0, \xi) = 0$, in contradiction with the noncharactericity of X_0.

The simplicity of the zeros implies that locally for $(x_0, \xi_0) \in T^*(X_0) \backslash 0$ the solutions of (5.1.3) are of the form $\xi^{(j)}(x_0, \xi_0)$, $j = 1, \ldots, \mu$ such that $(x_0, \xi_0) \to (x_0, \xi^{(j)}(x_0, \xi_0))$ is a homogeneous C^∞ mapping of degree 1 from $T^*(X_0) \backslash 0$ into $T^*(X)\big|_{X_0}$. In particular the number μ of solutions is locally constant in $T^*(X_0) \backslash 0$ and hence globally constant if X_0 is connected and $\dim X_0 \geq 2$, because then $T^*(X_0) \backslash 0$ is connected.

Definition 5.1.1. $P \in L^m(X)$ is called *strictly hyperbolic of multiplicity* μ *with respect to* X_0 if all bicharacteristic curves of P are transversal to X_0 and (5.1.3) has exactly μ solutions for every $(x_0, \xi_0) \in T^*(X_0) \setminus 0$.

If $Q_j \in L^{m_j}(X)$, $j = 1, \ldots, \nu$ are given we now want to construct operators $E_k: \mathcal{E}'(X_0) \to \mathcal{D}'(X)$, $k = 1, \ldots, \nu$, such that

$$(5.1.5) \qquad\qquad PE_k \equiv 0, \quad k = 1, \ldots, \nu$$

$$(5.1.6) \qquad\qquad \rho_0 Q_j E_k \equiv \delta_{jk} I, \quad j, k = 1, \ldots, \nu.$$

(We shall see in (5.1.13) that the number ν has to be equal to the number μ of zeros $\xi^{(j)}(X_0, \xi_0)$.)

Here $I = $ identity: $\mathcal{E}'(X_0) \to \mathcal{E}'(X_0)$, $A \equiv B$ means that $A - B$ is an integral operator with C^∞ kernel, and ρ_0 denotes restriction to X_0. The equations (5.1.5,6) imply that for any $g_k \in \mathcal{E}'(X_0)$, $k = 1, \ldots, \nu$ the distribution $u = \sum_k E_k g_k \in \mathcal{D}'(X)$ satisfies $Pu \in C^\infty(X)$ and $\rho_0 Q_j u = g_j \bmod C^\infty(X_0)$ for all $j = 1, \ldots, \nu$, so *modulo* C^∞ the operators E_k generate solutions of the Cauchy-problem $Pu = 0$, $\rho_0 Q_j u = g_j$.

Let us try

$$E_k \in I^{-m_k - 1/4}(X, X_0; C_0).$$

The order follows from the observation that $\rho_0 \in I^{1/4}$, $Q \in I^{m_k}$ and (5.1.6). To determine C_0 we observe that (5.1.5) more or less forces us to assume that $\tilde{p}_m = 0$ on C_0 if $\tilde{p}_m$ denotes the pullback of p_m to $T^*(X) \setminus 0 \times T^*(X_0) \setminus 0$ under the projection onto the first factor. So C_0 is invariant under the Hamilton field $H_{\tilde{p}_m}$ in view of (3.6.2) and the remark that the symplectic orthogonal complement of the tangent space of $\tilde{p}_m = 0$ is spanned by $H_{\tilde{p}_m}$ (all with respect to the symplectic form $\sigma = \sigma_{T^*(X)} - \sigma_{T^*(X_0)}$).

In view of (5.1.6) we need that $R_0 \times C_0$ intersects $T^*(X_0) \times \mathrm{diag}\, T^*(X) \times T^*(X_0)$ transversally and that $R_0 C_0 = \mathrm{diag}\, T^*(X_0) \setminus 0$, see Theorem 4.2.2. This leads to

$$
\begin{aligned}
C_0 = \{&(y, \eta), (x_0, \xi_0); (y, \eta) \text{ is on the bicharacteristic}\\
&\text{strip through some point } (x_0, \xi) \text{ such that}\\
&\xi|_{T_{x_0} X_0} = \xi_0 \text{ and } p_m(x_0, \xi) = 0\}
\end{aligned}
$$

(5.1.7)

and the condition

(5.1.8) Every bicharacteristic curve intersects X_0 at most once.

Now C_0 is the flow-out along the solution curves of $H_{\tilde{p}_m}$ of $\Delta_0 = N \times T^*(X_0) \cap R_0^{-1}$, where N is the characteristic hypersurface of P, defined in (5.1.1). The condition (5.1.4) implies that this intersection is transversal, so Δ_0 is a closed C^∞ conic manifold. It also implies that $H_{\tilde{p}_m}$ is transversal to Δ_0, so C_0 is an immersed Lagrange manifold in $T^*(X) \times T^*(X_0)$ in view of the arguments at the end of the proof of Theorem 3.6.2. (Note that Δ_0 is isotropic for σ and $\tilde{p}_m = 0$ on Δ_0.) Condition (5.1.4) implies also that $R_0 \times C_0$ is transversal to $T^*(X_0) \times \operatorname{diag} T^*(X) \times T^*(X_0)$.

The homogeneity of p_m implies that C_0 is conic. To see this, let a be a real nonzero homogeneous C^∞ function on $T^*(X) \times T^*(X_0) \setminus 0$ of degree $1 - m$. Then $q = a \cdot \tilde{p}_m$ is homogeneous of degree 1 and the H_q-flow commutes with multiplications by scalars, that is, maps cone axes into cone axes. $H_q = a \cdot H_{\tilde{p}_m}$ on $\tilde{p}_m = 0$, so the change from $\tilde{p}_m$ to q only involves a change of the time scale on the solution curves of the corresponding Hamilton fields. So C_0 is equal to the flow-out of the conic manifold Δ_0 along H_q, and therefore is conic. (This trick is also used in the regularization of the Kepler problem, see Moser [67].)

In order that the immersed C_0 is an embedded closed submanifold of $(T^*(X) \times T^*(X_0)) \setminus 0$, we need additional conditions on the global behavior of the bicharacteristic curves. Using the arguments of [22], Section 6.4, it is easily seen that the following conditions are necessary and sufficient:

(5.1.9)

    a) No bicharacteristic curve starting on X_0 stays in a compact subset of X,

    b) For every compact subset K_0 of X_0, K of X there is a compact subset K' of X such that if γ is an interval on a bicharacteristic curve with one end point in K_0 and the other in K, then $\gamma \subset K'$.

Finally the projection of C_0 into X (via $T^*(X)$) is a proper mapping if and only if

(5.1.10)   For every compact subset K of X there is a compact subset K_0 of X_0 such that every bicharacteristic curve starting in K only hits X_0 in K_0.

Note that (5.1.8,9,10) can always be satisfied if we replace X by a sufficiently small neighborhood of X_0, assuming that (5.1.4) holds.

Having settled the problems concerning C_0 we now turn to the actual construction of the operators $E_k \in I^{-m_k-1/4}(X, X_0; C_0)$. If e_k is the principal symbol of E_k then (5.1.5) leads to an equation of the form

(5.1.11) $$\frac{1}{i} \mathcal{L}_{H_{\tilde{p}_m}} e_k + \tilde{c} \cdot e_k = 0,$$

in view of Theorem 4.3.2. On the other hand, (5.1.6) leads in view of Theorem 4.2.2 to the algebraic equations

(5.1.12) $$\sum_{j=1}^{\mu} q_k(x_0, \xi^{(j)}(x_0, \xi_0)) \cdot r((x_0, \xi_0), (x_0, \xi^{(j)}(x_0, \xi_0)))$$

$$\times e_\ell((x_0, \xi^{(j)}(x_0, \xi_0)), (x_0, \xi_0)) = \delta_{k\ell}, \quad k, \ell = 1, \dots, \nu.$$

Here q_k and r denote the principal symbols of Q_k and ρ_0, respectively, and $\xi^{(j)}(x_0, \xi_0)$, $j = 1, \dots, \mu$, are the solutions of (5.1.3) depending smoothly on $(x_0, \xi_0) \in T^*(X_0) \setminus 0$. These equations have a unique solution if $\nu = \mu$ and

(5.1.13) The matrix $q_k(x_0, \xi^{(j)}(x_0, \xi_0))$, $k, j = 1, \dots, \mu$
is nonsingular for every $(x_0, \xi_0) \in T^*(X_0) \setminus 0$.

The solutions $e_\ell((x_0, \xi^{(j)}(x_0, \xi_0)), (x_0, \xi_0))$ can be treated as initial values on Δ_0 for the first-order differential equation (5.1.11) along the bicharacteristic strips and we obtain a unique solution e_k of (5.1.11,12) that is easily verified to belong to

$$S^{-m_k - \frac{1}{4} + \frac{1}{4}(2n-1)}(C_0, \Omega_{1/2} \otimes L_{C_0}).$$

(Use a trivialization of the line bundle over C_0 as discussed in Lemma 4.1.3 to reduce (5.1.11) to a differential equation for complex valued functions.)

Taking arbitrary $E_k^{(0)} \in I^{-m_k - 1/4}(X, X_0; C_0)$ with principal symbol e_k satisfying (5.1.11, 12), we have obtained that

$$PE_k^{(0)} \in I^{m - m_k - \frac{1}{4} - 1}(X, X_0; C_0)$$

and

$$\rho_0 Q_j E_k^{(0)} - \delta_{jk} I \in L^{-1}(X_0).$$

Recurrently solving differential equations along the bicharacteristic strips of the type (5.1.11) (but now inhomogeneous) and algebraic equations such as (5.1.12) for the initial data, we can determine the principal symbols of the operators $E_k^{(r)} \in I^{-m_k - \frac{1}{4} - r}(X, X_0; C_0)$ such that

$$P(e_k^{(0)} + \dots + E_k^{(r)}) \in I^{m - m_k - \frac{1}{4} - r - 1}(X, X_0; C_0)$$

and

$$\rho_0 Q_j (E_k^{(0)} + \cdots + E_k^{(r)}) - \delta_{jk} I \in L^{-r-1}(X_0)$$

for $r = 0, 1, 2, \ldots$. Taking for E_k an asymptotic sum of the $E_k^{(r)}$ (by taking for the local amplitude of E_k an asymptotic sum of the local amplitudes of the $E_k^{(r)}$), the construction of $E_k \in I^{-m_k-1/4}(X, X_0; C_0)$ satisfying (5.1.5,6) is complete. Summarizing we have proved the following theorem, which can be regarded as just a reformulation of the results of Lax [53] and Ludwig [56] in terms of the global theory of Fourier integral operators.

Theorem 5.1.2. *Let $P \in L^m(X)$ be a properly supported pseudodifferential operator with real and homogeneous principal symbol p_m, strictly hyperbolic of multiplicity μ with respect to the hypsersurface X_0 in X. If (5.1.8,9) hold and $Q_j \in L^{m_j}(X)$, $j = 1, \ldots, \mu$, have homogeneous principal symbols q_j satisfying (5.1.13), then there exist $E_k \in I^{-m_k-1/4}(X, X_0; C_0)$: $\mathcal{E}'(X_0) \to \mathcal{D}'(X)$ such that $PE_k \equiv 0$, $\rho_0 Q_j E_k \equiv \delta_{jk} I$. Here the homogeneous canonical relation C_0 is defined as in (5.1.7). If in addition (5.1.10) holds then we can choose these E_k to be continuous: $\mathcal{D}'(X_0) \to \mathcal{D}'(X)$.*

Remark. Using the parametrices of [22], Section 6.5, it can also be proved that the solution u of $Pu \in C^\infty(X)$, $\rho_0 Q_j u = g_j \bmod C^\infty(X_0)$ is uniquely determined modulo $C^\infty(X)$ and hence equal to $u = \sum E_k g_k \bmod C^\infty(X)$. It follows that $WF(u) \subset C_0 \circ \bigcup_{j=1}^{\mu} WF(g_j)$, so in particular the singular support of u is contained in the union of the bicharacteristic curves that are projections of bicharacteristic strips starting at points (x_0, ξ) such that $p_m(x_0, \xi) = 0$, $(x_0, \xi|_{T_{x_0} X_0}) \in WF(g_j)$. This is a precise formulation of the principle that "the singularities propagate along the bicharacteristic curves". Making only a microlocal construction of operators E_k, one can prove results on reflection of bicharacteristic curves at the boundary as in Nirenberg [69].

The following example shows that the bicharacteristic curves even may have singularities (turning points) without invalidating Theorem 5.1.2. In fact in the proof of Theorem 5.1.2 we did not see any effects of such turning points because the construction was made entirely in the cotangent bundle where the bicharacteristic strips are regular.

Example. $P = x_2 \left(\frac{\partial}{\partial x_1} \right)^2 + \left(\frac{\partial}{\partial x_2} \right)^2$ in $\mathbb{R}^2$ (Tricomi operator). The dotted line X_0 satisfies (5.1.8,9) but not (5.1.10). (The latter condition can be satisfied by choosing X = domain below the bicharacteristic curve γ.)

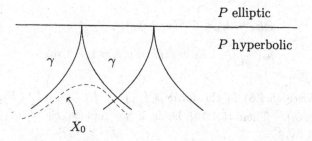

Instead of pursuing further results for general pseudodifferential operators P we now restrict attention to the case that P is a *differential* operator of order m on an open subset Y of $X \times \mathbb{R}$, X an n-dimensional manifold. Points in Y will be denoted by (x, t), $x \in X$, $t \in \mathbb{R}$. We assume that P satisfies the conditions of Theorem 5.1.2 with $\mu = m$ and X replaced by Y and X_0 by $X_s = (X \times \{s\}) \cap Y$ (the slight change in the notations should cause no trouble). Such operators are called *strictly hyperbolic* with respect to X_s. Following the presentation of Chazarain [16] and Chazarain and Piriou [17] we shall show that the construction of Theorem 5.1.2 can be used to obtain a unique solution $u \in C^\infty(Y)$ of the (inhomogeneous) exact Cauchy problem

$$
\begin{aligned}
Pu &= f \quad \text{in } Y \\
\rho_0 \left(\frac{\partial}{\partial t} \right)^{m-j} u &= g_j \quad \text{in } X_0, \quad j = 1, \ldots, m,
\end{aligned}
$$
(5.1.14)

for every $f \in C^\infty(Y)$, $g_j \in C^\infty(X_0)$. That is, $Q_j = \left(\frac{\partial}{\partial t} \right)^{m-j}$, $q_j = (i\tau)^{m-j}$, so (5.1.13) follows from the study of a Vandermonde determinant.

Compared with the classical proof using a priori L^2-estimates (due to Leray and Gårding; see Leray [55]) this "constructive" approach leads to much more detailed information about the operators E, E_k, which generate the solution $u = Ef + \sum_{k=1}^{m} E_k g_k$, see Theorem 5.1.6 below. On the other hand, the "energy estimates" are less sensitive to the geometrical conditions that we impose on the principal symbol, and can be generalized to cases where the construction of Theorem 5.1.2 apparently breaks down or leads to great complications.

For the necessity of hyperbolicity for the well-posedness of the Cauchy problem we refer to Lax [53], Section 2, with the modifications of Hörmander [44], Section 6.1, needed in the case of C^∞ coefficients.

Lemma 5.1.3. *The operators E_k in Theorem 5.1.2 can be chosen such that*

$$PE_k = R_k \in C^\infty(Y \times X_0),$$

(5.1.15)
$$\rho_0 \left(\frac{\partial}{\partial t} \right)^{m-j} E_k = \delta_{jk} I, \quad j, k = 1, \dots, m.$$

Proof. Write (5.1.6) in the form $\rho_0 \left(\frac{\partial}{\partial t} \right)^{m-j} E_k = \delta_{jk} I + R_{jk}$, $R_{jk} \in C^\infty(X_0 \times X_0)$. Then (5.1.15) holds if we replace $E_k = E_k(y, x_0) = E_k(x, t; x_0)$ by

$$E_k(x, t; x_0) - \sum_{j=1}^{m} \frac{t^{m-j}}{(m-j)!} R_{jk}(x, 0; x_0). \qquad \square$$

Lemma 5.1.4. *Let $Y = X \times \mathbb{R}$, suppose P is strictly hyperbolic with respect to $X_s = X \times \{s\}$ for every $s \in \mathbb{R}$. Then there exists for each $y \in Y$ a neighborhood U of y and a family of continuous mappings $E^{(s)} \colon C^\infty(Y) \to C^\infty(Y)$ depending smoothly on $s \in \mathbb{R}$, such that for all $f \in C^\infty(Y)$:*

$$PE^{(s)} f = f \ \text{in} \ U$$

$$\rho_s \left(\frac{\partial}{\partial t} \right)^{m-j} E^{(s)} f = 0 \ \text{in} \ X_s, \quad j = 1, \dots, m, \ s \in \mathbb{R}.$$

Here ρ_s is the restriction operator: $C^\infty(Y) \to C^\infty(X_s)$.

Proof. Replacing X_0 by X_s one obtains operators $E_k^{(s)}$ as in Lemma 5.1.3 with ρ_0 replaced by ρ_s and it is easily verified that they can be chosen such that $C^\infty(X) \to C^\infty(X \times \{s\}) \xrightarrow{E_k^{(s)}} C^\infty(Y)$ depends smoothly on s. Now write (Duhamel's principle):

(5.1.16)
$$(E^{(s)} f)(x, t) = \int_s^t (E_1^{(\tau)} \rho_\tau f)(x, t) \, d\tau.$$

Then it follows immediately that

$$(PE^{(s)} f)(x, t) = f(x, t) + (R^{(s)} f)(x, t)$$

$$\rho_s \left(\frac{\partial}{\partial t} \right)^{m-j} E^{(s)} f = 0, \quad j = 1, \dots, m$$

where

$$(R^{(s)} f)(x, t) = \int_s^t (PE_1^{(\tau)} \rho_\tau f)(x, t) \, d\tau.$$

Note that $PE_1^{(\tau)}$ has a smooth kernel.

Let $\psi \in C_0^\infty(Y)$ be equal to 1 on a neighborhood U of y, $0 \le \psi \le 1$. If supp ψ is sufficiently small then $\psi R^{(s)}$ will have norm $\le \rho < 1$ uniformly in s, when regarded as an operator in the Banach space of continuous functions on supp ψ with the supremum norm. It follows that $I + \psi P^{(s)}$ has a two-sided inverse depending smoothly on s. The proof is completed by replacing $E^{(s)}$ by $E^{(s)}(I + \psi R^{(s)})^{-1}\psi$. $\qquad\Box$

Lemma 5.1.5. *Let P be a strictly hyperbolic differential operator of order m on Y with respect to X_0. Then there is a neighborhood $\tilde{Y}$ of X_0 in Y, such that $u \in \mathcal{D}'(Y)$, $Pu = 0$ in Y, $\rho_0 \left(\frac{\partial}{\partial t}\right)^{m-j} u = 0$ in X_0 for $j = 1,\ldots,m$, implies $u = 0$ in $\tilde{Y}$.*

Proof. From Theorem 1.3.4 and Proposition 5.1.1 we see that we can multiply u by the Heaviside function $H(t)$ ($= 1$ for $t > 0$, $= 0$ for $t \le 0$), the result $\tilde{u} = Hu$ still being a solution of $P\tilde{u} = 0$ because of the vanishing of the Cauchy data. Let $d(x, x_0)$ denote the geodesic distance between x and x_0, with respect to some Riemann metric in X; fix $x_0 \in X$. Then there exists $\varepsilon_0 > 0$ such that for all $0 < \varepsilon < \varepsilon_0$, P is also strictly hyperbolic with respect to the hypersurface $S_\varepsilon = \{(x, \varepsilon^2 - d(x, x_0)^2);\ d(x, x_0) \le \varepsilon\}$.

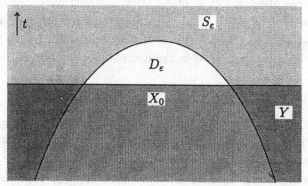

Now P is strictly hyperbolic if and only if its transposed ${}^t P$ is. Applying Lemma 5.1.4 we can find for any $g \in C^\infty(Y)$ a function $v \in C^\infty(Y)$ such that ${}^t P v = g$ in a neighborhood of the domain $D_\varepsilon = \{(x, t);\ 0 \le t \le \varepsilon^2 - d(x, x_0)^2\}$, with vanishing Cauchy data on S_ε. Using the uniqueness of the Cauchy problem on the level of formal power series, we see that all derivatives of v vanish on S_ε, so taking $v = 0$ for $t \ge \varepsilon^2 - d(x, x_0)^2$ leads to $\tilde{v} \in C^\infty(Y)$ such that ${}^t P \tilde{v} = \tilde{g}$ satisfies $\tilde{g} = g$ in $\{(x, t);\ t < \varepsilon^2 - d(x, x_0)^2\}$. It follows that $\langle \tilde{u}, \tilde{g} \rangle = \langle \tilde{u}, {}^t P \tilde{v} \rangle = \langle P\tilde{u}, \tilde{v} \rangle = 0$, and because $\tilde{g}$ is arbitrary in a neighborhood of D_ε in $\{(x, t);\ t < \varepsilon^2 - d(x, x_0)^2\}$ we find that $\tilde{u}$

vanishes there. Repeating the argument with $(1 - H)u$ instead of Hu leads to $u = 0$ in a neighborhood of x_0. Because $x_0 \in X_0$ is arbitrary, the lemma is proved.   □

Remark. The proof is somewhat simpler if we use the remark after Theorem 5.1.2 to obtain first that u is smooth. However, I wanted to make the proof independent of this reference to [22], in fact the result that we eventually obtain (Theorem 5.1.6) implies this regularity theorem and gives an alternative construction of a parametrix for P. Note also that if P has real analytic coefficients on a real analytic manifold, then it is sufficient that X_0 is noncharacteristic for P in order to construct v as above for a dense set of g; take g real analytic and use the theorem of Cauchy–Kowalewski. So in the case of analytic coefficients we obtain uniqueness for distribution solutions of the Cauchy problem under the weaker condition that X_0 is noncharacteristic for P (Theorem of Holmgren [39]).

A combination of Lemmas 5.1.3, 5.1.4, and 5.1.5 leads to a local existence and uniqueness theorem for the full Cauchy problem (5.1.14). Because of the local uniqueness, the local solutions automatically piece together in a neighborhood of all of X_0.

The proof of Lemma 5.15 actually gave the following conclusion. Let S be a hypersurface in Y such that P is strictly hyperbolic with respect to S, and at the same time equal to the boundary of a subset S_- of Y, such that

$$D = \{(x,t) \in S \cup S_-;\ t \geq 0\},$$

the domain between S and $t = 0$ is compact. Then $u = 0$ in a neighborhood of D, if $Pu = 0$ in a neighborhood of D and u has zero Cauchy data $\rho_0\left(\frac{\partial}{\partial t}\right)^{n-j}u$ in a neigborhood of the set of x such that $(x,0) \in S \cup S_-$.

Now let $y_0 = (x_0, t_0) \in Y$, $t_0 > 0$ be fixed. Taking all possible hypersurfaces S as above such that $y_0 \in D$, we see that $u = 0$ in a neighborhood of y_0, if $u \in \mathcal{D}'(Y)$, U_0 open in X_0 such that:

(a) $Pu = 0$, and $\rho_0 \left(\frac{\partial}{\partial t}\right)^{m-j} u = 0$ in U_0

(5.1.17)

(b) Every curve $y(t) = (x(t), t)$, $y(t_0) = y_0$, in Y such that $\frac{dy(t)}{dt} \in \Gamma^*(y(t))$ for all $0 \leq t \leq t_0$ hits U_0 for $t = 0$.

Here the cone $\Gamma^*(y) \subset T_y Y$ is defined as the convex hull of the downward side ($\delta t < 0$) of the set of tangent vectors to the bicharacteristic curves through y. (Because of the hyperbolicity no such vectors lie in the hyperplane $\delta t = 0$.) The set of points in Y that can be reached from y by curves $y(t)$ as in (b), $t \leq t_0$, will be called the *domain of dependence* $D(y)$. We also write $D_t(y) = D(y) \cap X_t$ and we have obtained that a solution of $Pu = f$ is uniquely determined in a neighborhood of $D_t(y)$ (for every $t \leq t_0$), as long as $D(y)$ between t and t_0 is compact. The method of sweeping out the domain by noncharacteristic surfaces to obtain the domain of dependence as outlined above is due to John [47]. In this way we have sketched a proof of:

Theorem 5.1.6. *Let P be a differential operator of order m on an open subset Y of $X \times \mathbb{R}$, strictly hyperbolic with respect to $X_s = (X \times \{s\}) \cap Y$ for every $s \in \mathbb{R}$. Assume moreover that the domain of dependence $D(y)$ between y and X_0 is compact for every $y \in Y$. Then the Cauchy problem (5.1.14) has for every $f \in C^\infty(Y)$, $g_j \equiv C^\infty(X_0)$ a unique solution $u = Ef + \sum_{j=1}^{m} E_j g_j$. The solution operators E, E_j are continuous: $C^\infty(Y) \to C^\infty(Y)$, respectively, $C^\infty(X_0) \to C^\infty(Y)$ and have the following properties:*

$$(5.1.18) \qquad \text{supp } E \subset \{(y, y') \in Y \times Y;\ y' \in D(y)\}$$

$$(5.1.19) \qquad \text{supp } E_j \subset \{(y, y_0) \in Y \times Y_0;\ y_0 \in D_0(y)\}$$

$$(5.1.20) \qquad E_j \in I^{j-m-1/4}(Y, Y_0; C_0) \qquad \text{(see (5.1.7))}$$

$$(5.1.21) \qquad WF'(E) \subset \text{diag } T^*(Y) \cup C \cup (C_0 \circ R_0),$$

where C is the set of all $((y, \eta), (y', \eta'))$ such that (y, η) and (y', η') are lying on the same bicharacteristic strip for P. On the different parts of the right-hand side in (5.1.21) the operator E can be identified as follows. If $A, B \in L^0(Y)$, $(WF(A) \times WF(B)) \cap (C \cup (C_0 \circ R_0)) = \emptyset$ then $AEB \in L^{-m}(Y)$, whereas $AEB \in I^{1-m-1/4}(Y, Y; C)$ if $(WF(A) \times WF(B)) \cap (\text{diag } T^*(Y) \cup (C_0 \circ R_0)) = \emptyset$.

Finally, if $B \in L^0(Y)$, $WF(B) \cap X_0^\perp = \emptyset$, then EB is continuous: $H_{\text{loc}}^s(Y) \to H_{\text{loc}}^{s+m-1}(Y)$ for all $s \in \mathbb{R}$.

Proof. The theorem is first proved in neighborhoods of the dependence domains $D(y)$, using induction with respect to t. All the time the E_j are modulo operators with C^∞ kernel equal to the operators constructed in Theorem 5.1.2 because the local solutions automatically piece together to global ones in view of the uniqueness. The statements about E can be proved by a closer examination of (5.1.16). Compare also with the parametrix E constructed in [22], Section 6.5.   □

Remarks. E, respectively, E_j can be continuously extended to $\{f \in \mathcal{D}'(Y); WF(f) \cap X_0^{\perp} = \emptyset\}$, resp., to $\mathcal{D}'(X_0)$, still solving the Cauchy problem for f, resp., g_j in these spaces. It follows from (5.1.20,21) that $(y, \eta) \in WF(u)$ only if $(y, \eta) \in WF(f)$, or $p_m(y, \eta) = 0$ and either $(y', \eta') \in WF(f)$ for some (y', η') on the bicharacteristic strip through (y, η) between X_0 and y or $(y_0', \xi') \in WF(g_j)$, $(y_0', \eta') \in$ on the bicharacteristic strip through (y, η), $\eta'|_{T_{y_0} X_0} = \xi'$. This is a refined version of the theorem on propagation of singularities due to Courant and Lax [19].

From (5.1.20) we obtain that E_j is continuous: $H_{\text{loc}}^s(X_0) \longrightarrow H_{\text{loc}}^{s+m-j}(Y)$ for every $s \in \mathbb{R}$, in view of Theorem 4.4.4 and that E is continuous: $H_{\text{loc}}^s(Y) \cap \mathcal{D}'X_0^{\perp}(Y) \to H_{\text{loc}}^{s+m-1}(Y)$ for every $s \in \mathbb{R}$.

Note also that

$$(5.1.22) \qquad \rho_t \left(\frac{\partial}{\partial t} \right)^{m-k} E_j \in I^{j-k}(X_t, X_0; C_{t_0}),$$

where $C_{t_0} = \{((y, \xi), (y_0, \xi_0)); \text{ there exists } \eta, \eta_0 \text{ such that } ((y, \eta), (y_0, \eta_0)) \in C, \eta|_{T_y X_t} = \xi, \eta_0|_{T_{y_0} X_0} = \xi_0\}$. C_{t_0} is the finite union of graphs of homogeneous canonical transformations: $T^*(X_0) \setminus 0 \to T^*(X_t) \setminus 0$, and the example of the wave operator $P = \Delta_x - (\partial/\partial t)^2$ shows that these are quite different from the identity and not even induced by a diffeomorphism $\kappa: X_0 \to X_t$. It follows in particular that this operator, mapping Cauchy data on X_0 of order $m - j$ to Cauchy data on X_t of order $m - k$, is continuous: $H_{\text{loc}}^s(X_0) \to H_{\text{loc}}^{s+k-j}(X_t)$ for all $s \in \mathbb{R}$. (Using only the H^s-continuity of the E_j we would have obtained continuity: $H^s \to H^{s+k-j-1/2}$.) A variant of this continuity result for operators with C^k coefficients, $k < \infty$, leads also to existence, uniqueness and regularity for nonlinear hyperbolic equations, see Fischer and Marsden [29], Section 2, and the literature cited there.

The construction of Theorem 5.1.2 meets serious complications if we want to apply it to mixed problems for hyperbolic equations in the presence of so-called "glancing rays," see Ludwig [58] for some formulas for the solutions of the wave equation in the exterior of a convex reflecting body. However these formulas, complicated as they are, do lead to a simple description of the propagation of singularities, see Morawetz and Ludwig [65].

We conclude this section by some remarks on the domains of dependence, which play such an important role in Theorem 5.1.6. Let $y \in Y$ and let the differential operator P of order m be strictly hyperbolic with respect to the plane $\langle \delta y, \nu \rangle = 0$, $\nu \in (T_y Y)^* \setminus \{0\}$. Let $\Gamma(y)$ be the component of ν in the complement of $N(y) = \{\eta \in (T_y Y)^* \setminus \{0\}; \ p_m(y, \eta) = 0\}$ in $(T_y Y)^* \setminus \{0\}$. Combining the results of Hörmander [44], Section 5.5, it follows that $\Gamma(y)$ is convex, its boundary is a component of the smooth hypersurface $N(y)$, and P is strictly hyperbolic with respect to the plane $\langle \delta y, \nu' \rangle = 0$ for any $\nu' \in \Gamma(y)$. It follows that the cone $\Gamma^*(y)$ defined in (5.1.17b) is equal to

$$(5.1.23) \qquad \Gamma^*(y) = \{\delta y \in T_y Y; \ \langle \delta y, \nu \rangle \leq 0 \ \text{ for all } \ \nu \in \Gamma(y)\}$$

or its antipodal. The smoothness of $\partial \Gamma(y)$ implies that $\Gamma^*(y)$ has no flat sides and that $\partial \Gamma^*(y)$ consists entirely of bicharacteristic tangent vectors. In other words, the map $\eta \mapsto d_\eta p_m(y, \eta)$ is surjective: $\partial \Gamma(y) \to \partial \Gamma^*(y)$.

Now assume that

$$(5.1.24) \qquad d_\eta^2 p_m(y, \eta) \ \text{ is nondegenerate for } \ \eta \in \partial \Gamma(y).$$

Then $\eta \to d_\eta p_m(y, \eta)$ is a diffeomorphism: $\partial \Gamma(y) \to \partial \Gamma^*(y)$, so in particular, $\partial \Gamma^*(y)$ is smooth. Using a variational principle (see Leray [55], Lemma 97.1, and Section 3.8 in these notes) it can be proved that each point of the boundary of $D(y)$ lies on a bicharacteristic curve issuing from y, so $\partial D(y)$ also consists entirely of bicharacteristic curves. If $\varepsilon > 0$ is sufficiently small, $y = (x, t_0)$ then the part of $\partial D(y)$ for $t_0 - \varepsilon < t \leq t_0$ is a smooth conoid with vertex at y. However, the global $\partial D(y)$ can develop singularities; an example is sketched below. (*See figure on following page.*)

From the figure we can see that a bicharacteristic curve on $\partial D_0(y)$ can leave $\partial D_0(y)$ by dipping into the interior of $D_0(y)$. (The variational principle mentioned above forbids leaving $D_0(y)$.)

Note that $\partial D(y)$ is the outermost component of the projection into Y of the Lagrange manifold Λ obtained by flowing out $N_{y_0} = \{(y_0, \eta) \in T^*(Y); \ p_m(y_0, \eta) = 0\}$ along H_{p_m} (in the negative t-direction).

For hyperbolic operators with constant coefficients still many more details are known, see Atiyah–Gårding [8] and the literature cited there.

The authors especially treat the problem of *lacunas*, which in its simplest form treats the question of whether $\partial D(y)$ is not only the singular support but even the support of the solution with Cauchy data = Dirac

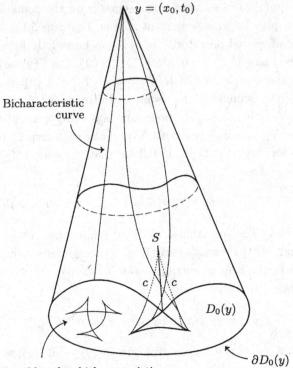

$y = (x_0, t_0)$

Bicharacteristic curve

S

c c

$D_0(y)$

$\partial D_0(y)$

Possible other bicharacteristics
in the case $m > 2$. These cones
can be singular from the start.

measure at y. Hadamard, in his classical book [38] on hyperbolic equations, calls this "Huygens' minor premise, paraphrasing [45]." Apart from the fact that what Hadamard calls "Huygens' major premise" does not play a role in [45], it should be remarked that Huygens states explicitly that behind the wave front there *is* a contribution of the "partial wavelets," which however is "infinitesimally small compared with the disturbance at the front" which is an envelope of the partial wavelets. In contemporary language, the solution has a singularity at the front $\partial D(y)$. Similar reasoning, with light rays instead of wave fronts, occurs in the explanation of Newton [68] of the rainbow, where it is concluded that singularly much light comes out of the water droplet at the maximal angle with the incident rays. So the determination of singularities was not uncommon in those days and in the case of the rainbow it is even more evident that "infinitesimally small compared with" did not imply "equal to zero" in such reasoning. Also

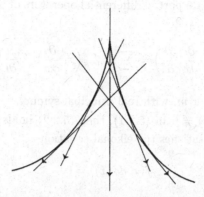

Pattern of the bicharacteristic curves in X. They have an envelope, forming a cusp that is the projection into X of the cusp-lines c of the "swallow's tail" in $\partial D_0(y)$. Such singularities correspond to caustics for the highly oscillatory asymptotic solutions for the reduced wave equation, issuing from x_0. See Section 5.2.

the theorem on propagation of singularities along the bicharacteristics in its microlocal (wave front) form can be seen in [45], keeping in mind, of course, that Huygens only gives a geometric theory of disturbances that is quite far from being an analysis of partial differential equations. See also [10], for a multidisciplinary view on Huygens' principle.

5.2. Oscillatory asymptotic solutions. Caustics

We return to the local asymptotic solutions

$$(5.2.1) \qquad u(x,\tau) = e^{i\tau\phi(x)}a(x,\tau), \qquad a(x,\tau) \sim \sum_{j=0}^{\infty} a_j(x)\tau^{\mu-j}$$

used to explain the validity of geometrical optics for wave mechanics as described in the Introduction. To fix ideas, we assume that $u(x,\tau)$ is an asymptotic solution, in an open subset U of an n-dimensional C^∞ manifold X, of a partial differential equation with large parameters, as follows:

$$(5.2.2) \qquad \sum_{k=0}^{m} P_{m-k}\left(x, \frac{\partial}{\partial x}\right) \cdot \left(\frac{\tau}{i}\right)^k u(x,\tau) = O(\tau^{-\ell}),$$

locally uniformly for $x \in U$, $\tau \to \infty$, all ℓ.

Here $P_{m-k}(x, \partial/\partial x)$ are partial differential operators of order $\leq m-k$ such that

$$(5.2.3) \qquad P\left(x, \frac{\partial}{\partial x}, \frac{\partial}{\partial t}\right) = \sum_{k=0}^{m} P_{m-k}\left(x, \frac{\partial}{\partial x}\right) \cdot \left(\frac{\partial}{\partial t}\right)^k$$

is an operator of order m, with real principal symbol.

If we want $a_0(x) \neq 0$ in (5.2.1), then (5.2.2) holds if and only if the phase function $\phi(x)$ satisfies the eikonal equation

$$(5.2.4) \qquad\qquad f(x, d\phi(x)) = 0$$

and the $a_j(x)$ satisfy a recurrent system of transport equations of the form

$$(5.2.5) \qquad d_\xi f(x, d\phi(x)) \cdot da_j(x) - g(x) \cdot a_j(x) = F_j(a_0, \ldots, a_{j-1})(x).$$

Here $f(x, \xi) = p_m(x; \xi, -1)$, $p_m = $ principal symbol of P of degree m. The factor $g(x)$ is a function (dependent on ϕ) that is closely related to the subprincipal symbol of P of degree $m - 1$. Finally $F_0 = 0$, that is, (5.2.5) starts with a homogeneous equation for a_0.

In view of Theorem 3.6.3 we always have local solutions ϕ of the nonlinear first-order partial differential equation (5.2.4) (nonlinear if $m > 1$), but in general ϕ cannot be extended to a global solution on X because of "turning vertical" of the flow-out along M_f of the graph of $d\phi$ (see the end of Section 3.6).

In order to obtain global asymptotic solutions of (5.2.2) perhaps the simplest way is to study the Fourier transform

$$(5.2.6) \qquad\qquad v(x, t) = \int_{-\infty}^{\infty} e^{-it\tau} u(x, \tau)\, d\tau$$

of $\tau \to u(x, \tau)$. Then (5.2.2) just means that

$$(5.2.7) \qquad\qquad P\left(x, \frac{\partial}{\partial x}, \frac{\partial}{\partial t}\right) v \in C^\infty(U \times \mathbb{R}).$$

Putting (5.2.6) in (5.2.1) shows that v is a Fourier integral distribution with amplitude $a(x, \tau)$, one frequency variable τ, and phase function $\tau \cdot [\phi(x) - t]$. So $v \in I^\nu(U \times \mathbb{R}, \phi^\perp)$ where $\nu = \mu + \frac{1}{2} - \frac{1}{4}(n+1)$ and

$$(5.2.8) \qquad \phi^\perp = \{(x, \phi(x), -\tau\, d\phi(x), t); \; x \in U, \; \tau \in \mathbb{R} \setminus \{0\}\}$$

is the normal bundle in $T^*(X \times \mathbb{R}) \setminus 0$ of the graph of ϕ in $U \times R$ (compare Lemma 3.7.5). This graph therefore is just the singular support of v in $U \times \mathbb{R}$.

Now the theory of Section 5.1 tells us exactly how the local solution $v \in I^\nu(U \times \mathbb{R}, \phi^\perp)$ of (5.2.7) extends to a global one. We have to flow out $\phi^\perp$ along the bicharacteristic strip of P to a global H_{p_m}-invariant conic Lagrange manifold C in $T^*(X \times \mathbb{R}) \setminus 0$, the symbol of the global solution $v \in I^\nu(X \times \mathbb{R}, C)$ of $Pv \in C^\infty(X \times \mathbb{R})$ is computed by recurrently solving first-order ordinary differential equations along the bicharacteristic strips as in the construction of the operators E_k in Theorem 5.1.2, taking them in $\phi^\perp$ equal to the corresponding symbols of the local solution v. (These equations along the bicharacteristic strips are the Fourier transform translation of the transport equations (5.2.5).) Uniqueness mod $C^\infty(X \times \mathbb{R}, C)$ of such a global solution follow also from Theorem 4.3.2. Finally we can get rid of the C^∞ error term by subtracting from v a C^∞ solution $\tilde{v}$ of $P\tilde{v} = Pv$, which exists if P is for example hyperbolic. Examples of global solutions $v \in I^\nu(X \times \mathbb{R}, C)$ of $Pv = 0$ are the $E_k \cdot \delta_{y_0}$, where the E_k are the solution operators in Theorem 5.1.6 and δ_{y_0} is the Dirac measure in X_{t_0} at $y_0 = (x_0, t_0)$. Thinking of any function in X_{t_0} as a superposition of δ-functions this example is considered as basic.

The inverse Fourier transform

$$(5.2.9) \qquad u(x, \tau) = (2\pi)^{-1} \int_{-\infty}^{\infty} e^{it} v(x, t)\, dt,$$

if it exists, then will represent a global solution of the *exact* equation $P(x, \partial/\partial x, \tau/i)u = 0$, which has the asymptotic expansion (5.2.1) in U if no bicharacteristic strip for P starting in $\phi^\perp$ returns from U. The existence of (5.2.9) depends on a sufficient decay of $v(x,t)$ as $|t| \to \infty$, locally uniformly in x, which can be a subtle question (see Lax and Phillips [54], Morawetz and Ludwig [64]). We can avoid such problems if the projection from $\pi(C) \subset X \times \mathbb{R}$ into X is a proper mapping, by cutting off v smoothly outside a closed neighborhood V of $\pi(C)$ where the projection $V \to X$ is still proper. Then (5.2.9) converges and defines a C^∞ function in (x,t), which is a global solution in X of the asymptotic equation (5.2.2). (The smoothness in x follows by testing with rapidly oscillating functions and observing that $\tau \neq 0$ on C. $\tau \neq 0$ on $\phi^\perp$ and τ is constant along the bicharacteristic curves of P because $\frac{\partial p_m}{\partial t} = 0$.)

Locally v is a finite sum of oscillatory integrals defined by a nondegenerate phase function $\phi(x,t,\theta)$ in $\mathbb{R}^{n+1} \times \mathbb{R}^N$ and a corresponding amplitude $a \in S^{\nu - \frac{1}{2}N + \frac{1}{4}(n+1)} = S^{\mu - \frac{1}{2}N + \frac{1}{2}}$ that is an asymptotic sum of homogeneous functions of θ. Hence $u(x,\tau)$ is a locally finite sum of integrals of the form

(5.2.10)
$$(2\pi)^{-1} \iint e^{i[\phi(x,t,\theta)+\tau t]} a(x,t,\theta)\, d\theta\, dt$$
$$= (2\pi)^{-1} \tau^N \iint e^{i\tau[\phi(x,t,\theta)+t]} a(x,t,\tau,\theta)\, d\theta\, dt,$$

which is an oscillatory integral as considered in Section 1.2, in $N+1$ variables. Note that the growth order of $\tau^N \cdot a(x,t,\tau,\theta)$ as $\tau \to \infty$ is equal to $N + \mu - \frac{1}{2}N + \frac{1}{2} = \mu + \frac{1}{2}(N+1)$.

Writing $\alpha = (\theta,t)$, $\psi(x,\alpha) = \phi(x,t,\theta) + t$, $k = N+1$, we see that $\psi(x,\alpha)$ is a nondegenerate phase function in the sense that

(5.2.11) $d_{(x,\alpha)}\, d_\alpha \psi$ has rank k when $d_\alpha \psi = 0$

if and only if

(5.2.12) C is transversal to $(x,t;\xi,\tau) \in T^*(X \times \mathbb{R});\ t = -1$.

This implies that

(5.2.13) $\Lambda = \{(x,\xi);\ \text{there is a } t \text{ such that } (x,t;\xi,-1) \in C\}$,

which is the projection into $T^*(X)$ of the intersection of C with $\{(x,t;\xi,\tau) \in T^*(X \times \mathbb{R});\ \tau = -1\}$, is an immersed C^∞ Lagrange manifold in $T^*(X)$, locally given by

(5.2.14) $\Lambda_\psi = \{(x,d_x\psi(x,\alpha));\ d_\alpha\psi(x,a) = 0\}$.

Note that if C is the flow-out along H_{p_m} of any C_0 satisfying (5.2.12) then C satisfies (5.2.12) because τ is constant along the bicharacteristics. In particular (5.2.12) holds for the flow-out C of $\phi^\perp$ and in this case Λ is just the flow-out in $T^*(X)$ along H_f of the graph of $d\phi$ in $T^*(X)$. As for conic Lagrange manifolds, every Λ_ψ defined by a function $\psi(x,\alpha)$ satisfying (5.2.11) is an immersed Lagrange manifold in $T^*(X)$ and conversely every immersed Lagrange manifold in $T^*(X)$ is locally of the form (5.2.14) for some ψ satisfying (5.2.11). We now summarize the above in a formal statement.

Definition 5.2.1. Let Λ be an immersed C^∞ Lagrange manifold in $T^*(X)$. An *oscillatory function $u(x,\tau)$ of order μ defined by Λ* is a locally finite (in X) sum of integrals of the form

$$(5.2.15) \qquad I(x,\tau) = \int e^{i\tau\psi(x,\alpha)} b(x,\alpha,\tau)\,d\alpha,$$

where $\alpha = (\alpha_1,\ldots,\alpha_k)$, $k \in \mathbb{N}$, ψ satisfies (5.2.11), Λ_ψ is a piece of Λ and finally

$$(5.2.16) \qquad b(x,\alpha,\tau) \sim \sum_{r=0}^{\infty} b_r(x,\alpha) \cdot \tau^{\mu+\frac{1}{2}k-r} \text{ and}$$

$b(x,\alpha,\tau)$ vanishes for α outside a fixed compact set in $\mathbb{R}^k$.

Proposition 5.2.2. *Let $i\colon \Lambda \to T^*(X)$ be the immersed Lagrange manifold in $T^*(X)$ by flowing out the graph of $d\phi$ along H_f, ϕ as in (5.2.1). More precisely, let Λ_0 be a hypersurface in $d\phi$ transversal to H_f, define $i(\lambda_0,t) = \Phi^t(\lambda_0)$, where Φ^t is the H_f-flow, $\lambda_0 \in \Lambda_0$, $t \in \mathbb{R}$, take $\Lambda = $ domain of definition of i in $\Lambda_0 \times \mathbb{R}$, assume that $d\phi \subset i(\Lambda)$. If $\pi \circ i\colon \Lambda \to X$ is proper and no solution curve of H_f starting in $d\varphi$ returns over U, then the local oscillatory asymptotic solution (5.2.1) of (5.2.2) can be extended to a global one of order μ defined by Λ. Two such global extensions differ by a function $\tilde{u}(x,\tau)$ that is rapidly decreasing $(\tilde{u}(x,\tau)) = O(\tau^{-\ell})$, $\tau \to \infty$, all ℓ).*

If every projection in X of an H_f-solution curve meets U then a stronger uniqueness result holds. If $\tilde{u}(x,\tau)$ satisfies (5.2.2) globally, if all x-derivatives of $\tilde{u}$ are bounded by a power of τ (locally uniformly in x) and $\tilde{u}$ and all its x-derivatives are rapidly decreasing in U, then $\tilde{u}$ and all its x-derivatives are rapidly decreasing in X.

The global calculus of oscillatory asymptotic functions defined by a Lagrange manifold was invented by Maslov [61] who assigns to every function a on Λ an oscillatory function u as in Definition 5.2.1 by means of a "canonical operator." Regarding a as the (principal) symbol of u, the solution u of (5.2.2) can be determined by recurrently solving the transport equations for a along the H_f-solution curves, corresponding to the transport equations on C for the principal symbol of the Fourier transform $v(x,t)$ of $u(x,\tau)$. See also Keller [48] who mentions the "multi-valued function $d\phi(x)$" (that is, the Lagrange manifold Λ) and the derivation of the phase shift when passing through a caustic from integral representations of the solutions obtained from an application of Green's theorem. This is more or less the program that we carried out in more detail here.

Note that the assumptions of Proposition 5.2.2 imply that $i^*\alpha$ is exact on Λ, here α = canonical 1-form in $T^*(X)$. Writing $i^*\alpha = d\psi$, we find $C = \{(x, \psi(x, \xi), \tau\xi, -\tau); \ (x, \xi) \in \Lambda, \ \tau \in \mathbb{R}\}$; here i is left out of the notation. Note also that the transversality of H_{p_m} to t = constant (on C) is equivalent to $d_\xi f(x, \xi) \cdot \xi \neq 0$ for all $(x, \xi) \in \Lambda$. All these assumptions are not really needed in all their strength and it is in fact very interesting to investigate more general cases.

Now we investigate the asymptotic behavior of the oscillatory integral. The only nonrapidly decreasing contributions come from the points (x_0, α_0) where $d_\alpha\psi(x_0, \alpha_0) = 0$. If $d_\alpha^2\psi(x_0, \alpha_0)$ is nondegenerate, then we obtain according to Section 1.2, an asymptotic expansion of the form (5.2.1) for the integral, with some other phase ϕ and amplitude a, but with the same growth order μ, so nothing new happens. However, if rank $d_\alpha^2\psi(x_0, \alpha_0) = r < k$ then only r of the α-variables can be integrated away with the method of stationary phase (see Lemma 2.3.5 for the details of this procedure) and we are left with an integral (5.2.15) with k replaced by $k - r$ and

$$(5.2.17) \qquad d_\alpha\psi(x_0, \alpha_0) = 0, \quad d_\alpha^2\psi(x_0, \alpha_0) = 0.$$

At such points the asymptotic value of the integral will be of higher order than τ^μ as $\tau \to \infty$ (although never of higher order than $\tau^{\mu + \frac{1}{2}(k-r)}$).

Now $k - r$ is just the dimension of $T_{(x_0, \xi_0)}(\Lambda_\psi) \cap T_{(x_0, \xi_0)}$ (fiber), $\xi_0 = d_x\psi(x_0, \xi_0)$ (see the remark before Lemma 2.3.5). So the points where exceptionally high intensity collects (where the light "burns," in optical terminology) are exactly those points where Λ turns vertical.

Definition 5.2.3. Let $i: \Lambda \to T^*(X)$ be an immersed Lagrange manifold in $T^*(X)$. The *caustic* $c(\Lambda)$ of Λ is the projection into X of the set $\Sigma(\Lambda)$ of points in $i(\Lambda)$ where i is not transversal to the fibers. At each point $x_0 \in X$ the *order of the caustic* is defined as the infimum $\kappa(x_0)$ of the numbers κ' such that $u(x, \tau) = O(\tau^{\mu + \kappa'})$ for $\tau \to \infty$, uniformly for x in a neighborhood of x_0, for any oscillatory function u of order μ defined by Λ.

Of course $\kappa(x_0) = 0$ for $x_0 \in \pi(\Lambda) \setminus c(\Lambda)$ and $\kappa(x_0) \leq k/2$ where k is the maximum of the dimensions of the intersections $T_{(x_0,\xi)}(\Lambda) \cap T_{(x_0,\xi)}$ (fiber) where $(x_0,\xi) \in \Lambda$. (Here i is left out of the notation, leading to a slight inconsistency.) However, it turns out that often (and in the generic case for $n \leq 5$), $0 < \kappa(x_0) < k/2$ for $x_0 \in c(\Lambda)$; see the list on the following page.

As in Section 1.2, the first step in obtaining an asymptotic expansion for (5.2.15) with a more general type of phase function ψ, is to try a change of variables in order to get ψ into some standard form. More precisely, suppose there exists a substitution of the form

$$(5.2.18) \qquad\qquad x = x(y), \qquad \alpha = \alpha(y,\beta)$$

which carries $\psi(x,\alpha)$ into $\chi(y,\beta)$ modulo a C^∞ function $\lambda(y)$ of y only, that is,

$$(5.2.19) \qquad\qquad \psi(x(y),\alpha(y,\beta)) + \lambda(y) = \chi(y,\beta).$$

Applying the substitution (5.2.18) to the integral (5.2.15) we obtain

$$(5.2.20) \qquad \tilde{I}(y,\tau) = I(x(y),\tau) = e^{-i\tau\lambda(y)} \int e^{i\tau\chi(y,\beta)} c(y,\beta,\tau)\, d\tau,$$

where $c(y,\beta,\tau) = b(x(y),\alpha(y,\beta),\tau) \cdot |\det d_\beta\alpha(y,\beta)|$. So apart from an additional oscillatory factor $e^{-i\tau\lambda(y)}$ and a transformation of variables $x = x(y)$ the integral is transformed into one with the new phase function $\chi(y,\beta)$.

It was recently observed independently by several authors (Guckenheimer [35], Arnol'd [3], [4], Guillemin and Schaeffer [36]) that (5.2.19) is just the definition of *equivalence* of ψ and χ, regarded as *unfoldings*, that is, as a family of functions of α, depending on parameters $x_1,\ldots,x_n$. The theory of unfoldings of singularities (a "singularity" of $\alpha \to \psi(x,\alpha)$ occurs when $d_\alpha\psi(x,\alpha) = 0$) was developed by Thom [79], and a rather complete theory can be given using the general theory of stability of C^∞ mappings of Mather [62]. Because of lack of time we shall not go into the details of this here, but only mention some of the most important results.

The first one is the complete classification of Arnol'd [4] of the so-called *simple* unfoldings, they all turn out to be locally equivalent to one of the following types:

Name	$\psi(x_1, \ldots, x_n; \alpha_1, \ldots, \alpha_k)$		κ
A_{m+1}	$\pm \alpha_1^{m+2} + Q + x_1\alpha_1 + \cdots + x_m\alpha_1^m,$	$n \geq m, k \geq 1$	$\frac{1}{2} - \frac{1}{m+2}$
$D_{m+1}^{\pm}$	$\alpha_1^2\alpha_2 \pm \alpha_2^m + Q + x_1\alpha_1 + x_2\alpha_2$ $+ \cdots + x_m\alpha_2^{m-1},$	$n \geq m, k \geq 2$	$\frac{1}{2} - \frac{1}{2m}$
E_6	$\alpha_1^3 \pm \alpha_2^4 + Q + x_1\alpha_1 + x_2\alpha_2 + x_3\alpha_2^2$ $+ x_4\alpha_1\alpha_2 + x_5\alpha_1\alpha_2^2$	$n \geq 5, k \geq 2$	$\frac{5}{12} = \frac{1}{2} - \frac{1}{12}$
E_7	$\alpha_1^3 + \alpha_1\alpha_2^3 + Q + x_1\alpha_1 + x_2\alpha_2 + x_3\alpha_2^2$ $+ x_4\alpha_2^3 + x_5\alpha_2^4 + x_6\alpha_1\alpha_2,$	$n \geq 6, k \geq 2$	$\frac{4}{9} = \frac{1}{2} - \frac{1}{18}$
E_8	$\alpha_1^3 + \alpha_2^5 + Q + x_1\alpha_1 + x_2\alpha_2 + x_3\alpha_2^2$ $+ x_4\alpha_2^3 + x_5\alpha_1\alpha_2 + x_6\alpha_1\alpha_2^2$ $+ x_7\alpha_1\alpha_2^3,$	$n \geq 7, k \geq 2$	$\frac{7}{15} = \frac{1}{2} - \frac{1}{30}$

Here

$$Q = \alpha_{k_0+1}^2 + \cdots + \alpha_{k_0+p}^2 - \alpha_{k_0+p+1}^2 - \cdots - \alpha_k^2, k_0 = 1 \text{ for } A_{m+1},$$
$$k_0 = 2 \text{ otherwise.}$$

The number κ in the last column is the order of the caustic at $x = 0$, these numbers can quite easily be determined using the semihomogeneity properties of the function ψ. Using a property of these ψ, called *infinitesimal stability* (Mather [62], II), one obtains for oscillatory integrals with these phase functions an asymptotic expansion of the form

(5.2.21)
$$I(x, \tau) \sim \sum_{r=0}^{\infty} \tau^{\mu + \frac{1}{2}k - \Sigma r_j - r} \left[\nu_r(x) \cdot F\left(\tau^{1-s_1}x_1, \ldots, \tau^{1-s_n}x_n\right) \right.$$
$$\left. + \sum_{\ell=1}^{n} \nu_{r,\ell}(x)\tau^{-s_\ell} \frac{\partial F}{\partial x_\ell}\left(\tau^{1-s_1}x_1, \ldots, \tau^{1-s_n}x_n\right) \right],$$

for $\tau \to \infty$, uniformly for x in a fixed neighborhood of $0 \in \mathbb{R}^n$. Here F is the "special function"

(5.2.22)
$$F(x_1, \ldots, x_n)$$
$$= \int \cdots \int e^{i\psi(x_1, \ldots, x_n; \alpha_1, \ldots, \alpha_k)} \, d\alpha_1 \cdots d\alpha_k,$$

ψ one of the standard unfoldings of type A_{m+1}, D_{m+1}, E_6, E_7, E_8. The integral (5.2.22) has to be interpreted by a shift of the path of integration into the complex domain, to make the integrand of order $O(e^{-\varepsilon_1|\alpha|^{\varepsilon_2}})$, $\varepsilon_1, \varepsilon_2 > 0$, and then defines an entire analytic function of $x \in \mathbb{C}^n$. Moreover it can be proved that F is bounded on R^n. The numbers s_ℓ, $\ell = 1, \ldots, n$ are defined by

$$(5.2.23) \qquad\qquad s_\ell = \sum r_j s_{j\ell},$$

where $\alpha_1^{s_{1\ell}} \cdot \alpha_2^{s_{2\ell}} \cdot \ldots \cdot \alpha_k^{s_{k\ell}}$ is the monomial $\partial \psi / \partial x_\ell$. Observe that $s_\ell < 1$ for all ℓ, this is in fact quite crucial in the proofs.

The simple unfoldings form an open and dense subset of $C^\infty(X \times A)$ for $n \leq 5$, and not so for $n \geq 6$, this is due to the appearance of "moduli." So generically for $n \leq 5$ every caustic is a locally finite superposition of caustics of the types A_{m+1}, D_{m+1} ($m \leq 5$), E_6. The simple unfoldings for $n \leq 4$ are the "elementary catastrophies" of Thom, they are called fold (A_2), cusp (A_3), swallow's tail (A_4), butterfly (A_5), elliptic umbilic (D_4^-), hyperbolic umbilic (D_4^+), parabolic umbilic (D_5). (Note that the $\pm$ signs in all cases except the D_{m+1} with m odd only decide between ψ and $-\psi$ and therefore are not very interesting.) Below we sketch the simple caustics for $n \leq 3$, the numbers are the orders on the different parts of the caustics.

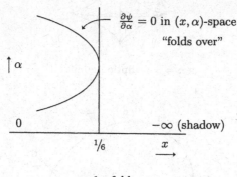

$n = 1$ fold

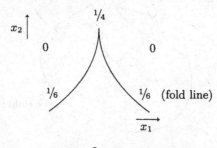

$n = 2$ cusp

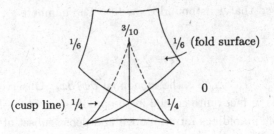

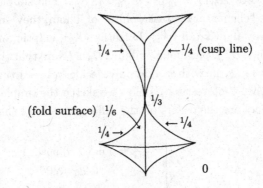

$n = 3$ swallow's tail

$n = 3$ elliptic umbilic

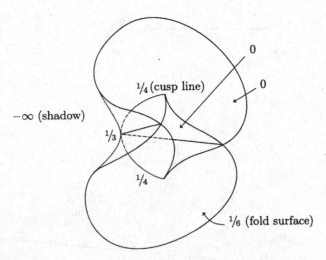

$n = 3$ hyperbolic umbilic

Note that for the swallow's tail and the hyperbolic umbilic the fold surface has self-intersections.

The uniform asymptotic expansion (5.2.21) was obtained for the fold by Ludwig [57], here the special function is

$$(5.2.24) \qquad Ai(x) = \int e^{i(\alpha^3 + x\alpha)} \, d\alpha \qquad \text{(Airy function)}$$

This is the famous integral introduced by Airy [2] in his study of the caustic that is responsible for the natural phenomenon of the rainbow. The idea of using the formula of infinitesimal stability in the proof can be found in Guillemin and Schaeffer [36].

For more details concerning stability of caustics, see the review article [21].

Originally I intended to also include some applications of Fourier integral operators as *similarity transformations*, but time has run out. Instead I refer to [22] where Fourier transformations are used to obtain similarity with the simple operators $\partial/\partial x_n$, respectively, $\partial/\partial x_n + i\,\partial/\partial x_{n-1}$ in the case where p is real, $dp \neq 0$ when $p = 0$, respectively, p is complex, $\{p, \bar{p}\} = 0$ and $d\,\mathrm{Re}\,p$, $d\,\mathrm{Im}\,p$ linearly independent when $p = 0$. In [73], [23], similarity with $\partial/\partial x_n + ix_n\,\partial/\partial x_{n-1}$ is obtained in the case that $\{p, \bar{p}\} \neq 0$ when $p = 0$. Fourier transformations are also used in order to simplify operators in the work of Egorov [24], [25] (to whom I ascribe this idea) and Nirenberg–Trèves [70].

References

[1] Abraham, R. and Marsden, J.E., *Foundations of Mechanics*, W.A. Benjamin, New York, Amsterdam, 1967.

[2] Airy, G.B., On the intensity of light in the neigborhood of a caustic, *Trans. Cambr. Phil. Soc.* (1838), 379–403.

[3] Arnol'd, V.I., Integrals of rapidly oscillating functions and singularities of the projections of Lagrange manifolds, *Funct. Anal. and Its Appl.* **6** 3 (1972), 61–62 (Russian).

[4] Arnol'd, V.I., Normal forms of functions with simple critical points, the Weyl groups A_k, D_k, E_k, and Lagrange immersions, *Funct. Anal. and its Appl.* **6** 4 (1972), 3–25 (Russian).

[5] Arnol'd, V.I., On the characteristic class entering in the quantization condition, *Func. Anal. Appl.* **1** (1967), 1–13.

[6] Arnol'd, V.I., Proof of a theorem of A.N. Kolmogorov on the invariance of quasi-periodic motions under small perturbations of the Hamiltonian, *Russ. Math. Surv.* **18** (1963), 9–36.

[7] Arnol'd, V.I. and Avez, A., *Ergodic Problems of Classical Mechanics*, W.A. Benjamin Inc., New York, Amsterdam, 1968.

[8] Atiyah, M.F., Bott, R., and Gårding, L., Lacunas for hyperbolic differential operators with constant coefficients, *Acta Mathematica* **124** (1970), 109–189.

[9] Birkhoff, G.D., Quantum mechanics and asymptotic series, *Bull. Amer. Math. Soc.* **39** (1933), 681–700.

[10] Blok, H., Ferwerda, H.A., Kuiken, H.K. (eds.), *Huygens' Principle 1690–1990, Theory and Applications*, North-Holland, Amsterdam, 1992.

[11] Born, M. and Wolf, E., *Principles of Optics*, Pergamon Press, London, 1959.

[12] Calderón, A.P. and Zygmund, A., Singular integral operators and differential operators, *Ann. J. Math.* **79** (1957), 901–921.

[13] Carathéodory, C., *Variationsrechnung und Partielle Differentialgleichungen*, Teubner, Berlin, 1935.

[14] Cartan, E., *Les Systèmes Différentiels Extérieurs et leurs Applications Geométriques*, Hermann, Paris, 1945.

[15] Cauchy, A., Théorie de la propagation des ondes (1815), *Oeuvres Complètes*, 1e Série, t. 1, Gauthier-Villars, Paris, 1882.

[16] Chazarain, J., Opérateurs hyperboliques à caractéristiques de multiplicité constante, Ann. Inst. Fourier **24** (1974), 173–202.

[17] Chazarain, J. et Piriou, A., *Introduction à la Théorie des Équations aux Dérivées Partielles Linéares*, Gauthier-Villars, Paris, 1981.

[18] Coddington, E.A. and Levinson, N., *Theory of Ordinary Differential Equations*, McGraw-Hill, New York, 1955.

[19] Courant, R. and Lax, P.D., The propagation of discontinuities in wave motion, *Proc. Nat. Acad. Sci. U.S.A.* **42** (1956), 872–876.

[20] de Rham, G., *Variétes Differentiables*, Hermann, Paris, 1955.

[21] Duistermaat, J.J., Oscillatory integrals, Lagrange immersions and unfoldings of singularities, *Comm. Pure Appl. Math.* **27** (1974), 207–281.

[22] Duistermaat, J.J. and Hörmander, L., Fourier integral operators II, *Acta Math.* **128** (1972), 183–269.

[23] Duistermaat, J.J. and Sjöstrand, J., A global construction for pseudo-differential operators with noninvolutive characteristics, *Inv. Math.*, **20** (1973), 209–225.

[24] Egorov, Yu.V., On canonical transformations of pseudo-differential operators, *Usp. Mat. Nauk* **25** (1969), 235–236.

[25] Egorov, Yu.V., On nondegenerate hypoelliptic pseudo-differential operators, *Sov. Math. Dokl.* **10** (1969), 697–699.

[26] Ehrenpreis, L., Solutions of some problems of division I, *Ann. J. Math.* **76** (1954), 883–903.

[27] Engel, F. and Faber, K., *Die Liesche Theorie der Partiellen Differentialgleichungen Erster Ordnung*, Teubner, Berlin, 1935.

[28] Erdelyi, A., *Asymptotic Expansions*, Dover Publications, 1956.

[29] Fischer, A.E. and Marsden, J.E., General relativity, partial differential equations and dynamical systems, Proc. Symp. Pure Math. AMS Summer Inst. on PDE, Berkeley, 1971.

[30] Fuks, D.B., On the characteristic classes of Maslov–Arnol'd, *Dokl. Akad. Nauk SSSR* **178** (1968), 303–306.

[31] Gabor, A., Remarks on the wave front set of a distribution, *Trans. Amer. Math. Soc.* **170** (1972), 239–244.

[32] Gårding, L., Kotake, T., and Leray, J. Uniformisation et développement asymptotique de la solution du problème de Cauchy linéaire, à données holomorphes; analogie avec la théorie des ondes asymptotiques et approchées (Problème de Cauchy I bis et VI), *Bull. Soc. Math. France* **92** (1961), 263–361.

[33] Godbillon, C., *Géometrie Differentielle et Mécanique Analytique*, Hermann, Paris, 1969.

[34] Grassmann, M., Die Ausdehnungslehre von 1844, 2 Kap., *Gesammelte Math. u. Phys. Werke*, Chelsea, New York, 1969.

[35] Guckenheimer, J., *Catastrophies and Partial Differential Equations, Annales de l'Institut Fourier*, **23** (1973), 31–59.

[36] Guillemin, V. and Schaeffer, D., Remarks on a paper of D. Ludwig, *Bull. A.M.S.* **79**, 2 (1973), 382–385.

[37] Gunning, R.C., *Lectures on Riemann Surfaces*, Princeton, 1966.

[38] Hadamard, J., *Le problème de Cauchy et les équations aux dérivées partielles linéaires hyperboliques*, Reprint of the 1923 publication, Dover, New York, 1952.

[39] Holmgren, E., Über Systeme von linearen partiellen Differentialgleichungen, *Öfversigt af Kongl. Vetenskaps-Akad. Förh.* **58** (1901), 91–105.

[40] Hörmander, L., Fourier integral operators I, *Acta Math.* **127** (1971), 79–183.

[41] Hörmander, L., Pseudo-differential operators and hypoelliptic equations, *AMS Symp. Pure Math.* **10** (1966), Singular Integral Operators, 138–183.

[42] Hörmander, L., Pseudo-differential operators, *Comm. Pure Appl. Math.* **18** (1965), 501–517.

[42'] Hörmander, L., *The Analysis of Linear Partial Differential Operators* I–IV, Springer-Verlag, Berlin, 1985. Vol: I: second edition, 1990.

[43] Hörmander, L., Uniqueness theorems and wave front sets for solutions of linear differential equations with analytic coefficients, *Comm. Pure Appl. Math.* **24** (1971), 671–704.

[44] Hörmander, L., *Linear Partial Differential Equations*, Springer-Verlag, 2nd ed., 1964.

[45] Huygens, C., *Treatise on Light (1678)*, Reprint of the English translation of 1690, Dover Publications, New York, 1962.

[46] Jacobi, C.G.J., *Vorlesungen über Dynamik (1847/48)*, A. Clebsch, Berlin, 1866.

[47] John, F., On linear partial differential equations with analytic coefficients: unique continuation of data, *Comm. Pure Appl. Math.* **2** (1949), 209–254.

[48] Keller, J.B., Corrected Bohr–Sommerfeld quantum conditions for non-separable systems, *Ann. of Physics* **4** (1958), 180–188.

[49] Kline, M. and Kay, I.W., *Electromagnetic Theory and Geometrical Optics*, Interscience (Wiley) Publ., 1964.

[50] Kohn, J.J. and Nirenberg, L., On the algebra of pseudodifferential operators, *Comm. Pure Appl. Math.* **18** (1965), 269–305.

[51] Lagrange, J.-L., *Mécanique Analytique*, reprint of 1811 edition, Albert Blanchard, Paris, 1965.

[52] Landau, E., Einige Ungleichungen für zweimal differentiierbare Funktionen, *Proc. London Math. Soc.* **13** (1914), 43–49.

[53] Lax, P.D., Asymptotic solutions of oscillatory initial value problems, *Duke Math. J.* **24**, (1957), 627–646.

[54] Lax, P.D. and Phillips, R.S. *Scattering Theory*, Academic Press, New York, 1967.

[55] Leray, J., *Hyperbolic Differential Equations*, Lecture Notes, The Institute for Advanced Study, Princeton, NJ, 1952.

[56] Ludwig, D., Exact and asymptotic solutions of the Cauchy problem, *Comm. Pure Appl. Math.* **13** (1960), 473–508.

[57] Ludwig, D., Uniform asymptotic expansion at a caustic, *Comm. Pure Appl. Math.* **19** (1966), 215–250.

[58] Ludwig, D., Uniform asymptotic expansions of the field scattered by a convex object at high frequencies, *Comm. Pure Appl. Math.* **16** (1963), 477–486.

[59] Luneburg, R.K., *Propagation of Electromagnetic Waves*, Lecture Notes, New York University, 1948.

[60] Malgrange, B., Existence et approximation des solutions des équations aux dérivées partielles et des équations de convolution, *Ann. Inst. Fourier*, Grenoble **6** (1955), 271–355.

[61] Maslov, V.P., *Perturbation Theory and Asymptotic Methods*, Moshov, Gozd. Univ., 1965 (Russian). French translation: *Théorie des Perturbations et Méthodes Asymptotiques*. Dunod, Gauthiers-Villars, Paris, 1972.

[62] Mather, J.N., *Stability of C^∞ mappings*. I. The division theorem, *Ann. of Math.* **87** (1968), 89–104. II. Infinitesimal stability implies stability, *Ann. of Math.* **89** (1969), 254–291. III. Finitely determined map-germs, *Publ. Math. I.H.E.S.* **35** (1968), 127–156. IV. Classification of stable germs by R-algebras, *Publ. Math. I.H.E.S.* **37** (1970), 223–248. V. Transversality, *Adv. in Math.* **4** (1970), 301–336. VI. The nice dimensions, pp. 207–253 in Proc. Liverpool Singularities I, *Lecture Notes in Mathematics* **192** (edited by C.T.C. Wall), Springer-Verlag, 1971.

[63] Maxwell, J.C., *A Treatise on Electricity and Magnetism*, Dover Publ., New York, 1873.

[64] Morawetz, C.S. and Ludwig, D., An inequality for the reduced wave operator and the justification of geometrical optics, *Comm. Pure Appl. Math.* **21** (1968), 187–203.

[65] Morawetz, C.S. and Ludwig, D., The generalized Huygens' principle for reflecting bodies, *Comm. Pure Appl. Math.* **22** (1969), 189–205.

[66] Moser, J., A rapidly convergent iteration method and nonlinear partial differential equations I, II, *Ann. Sc. Norm. Sup. Pisa*, **20** (1966), 265–315, 499–535.

[67] Moser, J., Regularization of Kepler's problem and the averaging method on a manifold, *Comm. Pure Appl. Math.* **23** (1970), 609–636.

[68] Newton, I., *Opticles*, Reprint of the 1730 edition, Dover Publications, New York, 1952.

[69] Nirenberg, L., *Lectures on Partial Differential Equations*, Proc. Reg. Conf. at Texas Tech., May 1972, Conf. Board of the A.M.S.

[70] Nirenberg, L. and Treves, F., On local solvability of linear partial differential equations, *Comm. Pure Appl. Math.* **23** (1970), 1–38 and 459–510.

[71] Sato, M., Hyperfunctions and partial differential equations, Proc. 2nd Conf. on Functional Anal., Tokyo, (1969), 91–94.

[72] Sato, M., Regularity of hyperfunction solutions of partial differential equations, pp. 785–794 in vol. 2 of *Actes Congr. Int. Math.*, Nice, 1970.

[73] Sato, M., Kawai, T., and Kashiwara, M., Microfunctions and pseudo-differential equations, to appear in Proc. Katata Conf. 1971, Part II, Springer Lecture Notes.

[74] Schwartz, L., Séminaire Schwartz 1 (1953–54), Produits tensoriels topologiques etc., Exposition No. 11.

[75] Shih, Weishu, On the symbol of a pseudo-differential operator, *Bull. Am. Math. Soc.* **74** (1968), 657–659.

[76] Sjöstrand, J., Operators of principal type with interior boundary conditions, *Acta Math.* **130** (1973), 1–51.

[77] Sommerfeld, A. and Runge, J., Anwendung der Vektorrechnung auf die Grundlagen der geometrischen Optik, *Ann. Phys.* **35** (1911), 277–298.

[78] Souriau, J.M., *Structure des Systèmes Dynamiques*, Dunod, Paris, 1970.

[79] Thom, R., *Modèles Mathématiques de la Morphogenèse*, Chapter 3: Théorie du déploiement universel, I.H.E.S., 1971.

[80] Thom, R., *Structural Stability and Morphogenesis*, W.A. Benjamin, Reading, Mass., 1972

[81] Weinstein, A., Symplectic manifolds and their Lagrangian submanifolds, *Adv. in Math.* **6** (1971), 329–346.

[82] Whittaker, E., *A Treatise on the Analytical Dynamics of Particles and Rigid Bodies*, 4th Ed., Cambridge Univ. Press, Cambridge, 1959.